Running Your Own
Driving School

GW00492886

Contents

Introduction

This book is written for the experienced driver wishing to qualify as a professional driving instructor and those wanting to establish their own driving school. Compared with most other businesses it is relatively easy to start up your own driving school. There are various reasons for this:

- Driving instruction is a large and well established market.
- The cost of training is fairly low.
- The period of training is fairly short.
- Driving instruction is an ideal cash flow business; ie, it produces immediate cash payments or payment in advance for courses.
- Relatively small capital outlay is required.

Various full- and part-time business and employment opportunities providing security and good prospects exist within the driving school industry for both men and women prepared to acquire the appropriate skills. Although the profession is still dominated by men, more women are beginning to enter it. This is having a beneficial effect since they are mostly conscientious and make excellent instructors. It is perhaps significant that the chairmen of the two largest national associations after the Driving Instructors' Association (DIA) are women.

In addition to providing traditional lessons for learners, the driving school industry supplies a wide range of other associated instructional, vocational and recreational services, which range from pre-driver courses and traffic education through PSV (passenger service vehicle) and HGV (heavy goods vehicle) courses to advanced and high performance driver instruction.

Driving instructors provide an essential service to around 1 million new drivers each year and make a significant contribution to the reduction of accidents. This helps to ensure our record is maintained as one of the safest driving nations of the world, despite the density of traffic on Britain's roads compared with most other advanced countries.

About 20,000 driving instructors are employed, either full or part time, in a well established market with an estimated annual turnover of £250 to £300 million. The industry provides a fundamental core of expertise in road safety and is a valuable source of future specialists in other areas of traffic education: research, development and driver testing.

The industry is able to provide employment, security and business opportunities for those prepared to acquire the appropriate skills. Each year around 1500 to 2000 new instructors are required to replace those who have moved on or retired.

Driving qualifications

The primary qualifications required by these new entrants into the industry are that:

- They have held a full driving licence for at least four out of the six years preceding the date of application.
- They have not been under a disqualification order for the last four years.
- They are fit and proper persons to have their names entered in the Department of Transport Register.
- They pass the qualifying examinations.

Personal qualities of driving instructors

It is a difficult, if not impossible task, to predict with any degree of certainty, what type of person is most suited to giving driving instruction, or to determine what previous experience and background they should have. The factors are so complex they are almost impossible to unravel. In any event, the student drivers on the receiving end of the instruction are so diverse in personality, beliefs, likes, dislikes etc, it would be unwise to attempt to identify one particular stereotyped instructor who would be perfectly suited to all.

However, some qualities must be common to all instructors. First of all, a proper and genuine concern for the student and his aspirations is necessary. Second, the instructor must be able to communicate in a manner which the student will understand and respect. Third, the instructor must have a thorough knowledge of the rules, procedures and skills involved and be able to demonstrate them. The instructor's attitude to driving and associated subjects is extremely important. Finally, the instructor must be totally aware of the road and traffic environment and of his student's needs at all times.

In his work, the driving instructor will meet all kinds of people from all walks of life and he will need to vary his approach according to their different personalities, attitudes, skills, abilities and aptitudes. It has been said that a driving instructor must be parent, guardian, nurse, teacher, doctor and psychiatrist to the student driver. It is certainly true that, in addition to being totally aware of traffic conditions, the instructor must plan routes to avoid or incorporate specific conditions according to the student's ability. He must also advise, show, explain, encourage, assess and protect students in order that they can learn effectively and safely.

Teaching driving requires patience, understanding, tact, firmness and a whole lot more because new drivers often require a tremendous amount of emotional support, particularly in times of difficulty. This calls for both physical and mental stamina. Persons teaching driving must be adequately practised in all the skills involved. This means they must themselves be skilful drivers with a thorough knowledge of the task and have the correct attitudes towards driving and the principles of safety it involves.

The qualities required of good instructors are very similar to those required of drivers and are identified in the Department of Transport manual *Driving* as follows:

Responsibility. A good driver will have a proper concern for his own safety and that of his passengers and other road users. A good instructor must have a proper concern for the well-being and future safety of his clients.

Concentration is part of the responsibility of driving a motor vehicle. A moment's distraction from the road and traffic environment can result in disaster. Instructors must not only concentrate on the road environment, but also upon how the student is responding to it.

Anticipation involves the driver's ability to protect himself from the actions of others and predict subsequent events.

Driving involves constant two-way communication between road users achieved through a complex system of signals and cues which occur at both conscious and subconscious levels. Good drivers are able to both give and interpret these signals and cues correctly, quickly and consistently.

An instructor must not only possess these skills as a driver but also be able to predict how a student is going to respond to these and other traffic situations. In addition the ability to communicate is one of the teacher's main assets. This is also a two-way system of communication between instructor and pupil. In simple terms, the driving instructor must be able to:

- Communicate ideas, facts, information and opinions to the learner.
- Provide the learner with appropriate opportunities for practice.
- Assess the learner's needs and performance in the skills being practised.

All of this should be at a level appropriate to the student's ability.

Patience. A driver who loses his temper because someone else makes a mistake, or because he is held up in traffic, is a potential source of danger to himself and others. An instructor who loses his temper because the student makes an error or is experiencing difficulty with a particular task, will destroy the student's confidence and cause even more errors and deficiencies. A good instructor will be sensitive to the student's problems and needs and will help to build confidence and improve the student's ability.

Confidence. Confidence is part of the driver's attitude to driving which usually increases with experience. Good drivers learn to recognise their limitations and avoid over-confidence which can lead to carelessness and mistakes.

Good drivers learn about the vehicle they are driving, which helps them to understand its servicing and maintenance needs. Although a detailed technical knowledge is not needed, a driving instructor will acquire a thorough knowledge of the laws relating to vehicle roadworthiness and a sufficient understanding of its operating principles to be able to answer questions posed by new drivers.

While it is extremely doubtful whether the perfect driver exists at all, it is something every driving instructor should seek to become. A good instructor will take pride in being an example to others. A perfect driver will at all times be in absolute harmony with both the vehicle and the road environment. This means that he or she will always be in the correct position travelling at an appropriate and safe speed with a suitable gear engaged to satisfy both the needs of the vehicle and the road and traffic conditions.

Driving at this level requires intense concentration and highly developed visual search and hazard recognition, skills which enable the perfect driver to minimise the risks and consequences of the actions and deficiencies of other road users. Perfection requires a thorough up-to-date knowledge of rules and procedures. Equally important is the driver's attitude to the vehicle, environmental factors, other road users, and other risk factors.

All the qualities so far discussed depend very much upon obtaining the correct instruction from the start.

The same is true for both drivers and driving instructors. Unprofessional teachers can ruin the confidence of new instructors as well as new drivers.

A properly trained and practised instructor should have confidence in himself, what he is teaching and how he is teaching it. He will also have an open mind and be prepared to change, adapt and adopt the ideas of others. There are no short cuts in learning to drive and the same applies to learning to become an instructor, as many have discovered through bitter experience.

The move to self-employment

Over the past twenty years there has been a continuous drift of driving instructors to the ranks of the self-employed. Many made this change only after careful consideration, while others had rather less idea of what was involved.

With current levels of unemployment, it is not surprising that more and more people are turning to self-employment. The ever-present threat of redundancies and financially crippling strikes may sometimes make self-employment seem a tame option, and certainly the gap between the security of the employee working for a large multinational company and the self-employed person working from home appears to have narrowed in recent years.

There is one main difference: the self-employed person has more control over his future than the employee and if he fails in business he has no one to blame but himself. In considering this change of employment you should recognise that there are few jobs where such an intense level of personal contact with clients prevails as in driving instruction or where more personal discipline is involved. An employee has discipline imposed on him — a self-employed person must discipline himself and not everyone is capable of doing this.

The skills required to be a successful businessman are considerably more complex than those required by the good instructor and becoming self-employed can cause a temporary lowering of one's standard of living. If all does not go as planned it can even drop permanently.

'Going it alone' can be a long, hard struggle, particularly in the absence of the moral support essential from close relatives. Before taking the plunge, the full implications should be thoroughly discussed with them and this will also assist in establishing their enthusiasm and the amount of help and encouragement likely to be forthcoming.

Although being self-employed provides a greater degree of independence than is ever likely to be achieved while working for an employer, anyone contemplating the change must also recognise that insecurity is an inescapable condition. Any investment in a new business venture contains an element of risk. Do not over-commit yourself and, as far as possible, try to finance it from your own personal resources.

Driving instructors usually measure their success in two ways—their personal achievements as an instructor and their financial success. These two things should not be confused; they are distinctly separate and success as a driving instructor will not automatically lead to business or financial achievement. Nor is business or financial success totally dependent upon being a good instructor, though it is an obvious advantage from a personal point of view. Being successful depends upon what the individual expects from life and whether he achieves those personal goals.

Business success may mean different things to different people. For example, you may consider yourself to be a success if you own a large company which employs numerous instructors, but which is not particularly profitable. On the other hand, you may consider that success means the financial rewards of a smaller but more profitable concern. Although success is a matter of personal judgement relating to the aspirations of the individual, it is usually associated with finance, material possessions and personal satisfaction. If you want to be a financial success then, in the words of J Paul Getty, 'Almost without exception there is only one way to make a great deal of money—and that is to own your own business.'

Comparison between employee status and self-employment

	Employee	*Self-employed*
Holidays	Holidays with pay.	Holidays without pay.
Injury	State industrial injury benefits.	No state industrial injury benefits.
Income tax	Schedule E tax PAYE. Limited expenses can be claimed.	Schedule D tax, paid twice yearly. Wider range of permissible expenses.
Liabilities	Employee and employer liable for injuries due to employee's negligence.	Self-employed person liable for own negligence.
National Insurance contributions	Paid jointly by employer and employee at higher rate.	Paid wholly by self-employed person at lower rate, on a weekly/monthly basis and further contributions made with income tax payments.
Notice	Entitled to the minimum laid down by law and the contract.	Not applicable.
Pension	Company or state scheme.	Tax relief on pension and annuity scheme payments.
Redundancy	Usually eligible for the state redundancy scheme.	Not eligible for the state scheme.
Sickness	Eligible for state sickness benefits and full or part wage payments during illness.	Eligible for the state sickness scheme. No payments usually made in lieu of wages.
Unemployment	Usually eligible for the state unemployment benefit scheme.	Not entitled to the state unemployment benefit scheme.

Under the Social Security Act 1973 self-employed persons are treated differently from employees for the purposes of income tax and National insurance payments.

Checklist for a prospective driving school owner

Before making any decision regarding your professional status, thought should be given to the following:

- Look realistically at the problems and risks involved. (It is important that anyone contemplating his own driving school should be a completely competent instructor.)
- Find out what competition exists in your area.
- Make sure you have calculated your figures correctly.
- Ensure you can support yourself financially long enough to establish the business (it is likely to take up to six months).
- Be prepared to obtain professional advice and assistance.
- Once you have made up your mind, commit yourself totally to the business.

During the initial period of getting yourself known, it may be prudent to remain in full-time employment, or alternatively find some part-time work for the first few months, until the business is able to sustain your living standards at a comfortable level.

If you would like to start up a driving school as a means of extending the services you are already providing within an existing business and intend to employ instructors rather than teach yourself, as a security measure it would be wise to qualify yourself, or have a senior member of your staff qualify, so that you are more able to control and monitor the quality of service you are providing.

Chapter 1

The Business and its Legal Structure

If you have been offered an opportunity or are considering the possibility of purchasing an already established driving school, you should not make a decision to buy before taking into account all the implications and researching its background and reputation. Establish very clearly why the school is for sale and if possible check the reasons given (owner's retirement is often cited).

While there are obvious advantages, such as the existing clients and the inflow of new ones, it may not work out as planned, particularly if you are only purchasing the goodwill of the instructor or proprietor of a small school. You must always assume that the seller is asking too high a price for the goodwill which may not materialise; for instance some of the customers may not wish to continue with a new instructor.

Whether it is a good buy or not will depend largely on the asking price and professional advice should be sought before any commitment is made. In some instances it may be far better to build up your own business from scratch.

In fact, driving schools rarely come on to the market. When they do, they are generally associated with the purchase of property, the value of which should be looked at in its own right in order to determine how much its price may have been inflated for the goodwill of the school.

Goodwill is a purchasable but intangible asset which, particularly in the case of a driving school, is difficult to value. To pay more than £1000 for the goodwill of a well-established and reputable one-car school would be a doubtful proposition and it is difficult to give reliable advice in this matter. It should be clearly understood that this is a maximum figure which in many cases will be too high.

Goodwill in a business which has only been running for a few years (up to eight or nine, say), may be very short-lived and possibly not worth more than £500 or so. On the other hand, a long-established business of, say, 15 to 20 years' standing, where the owner is retiring, might be a more realistic proposition. It should be recognised that goodwill is normally very short-lived and pertains to the person who has built it up. If you can't recover the costs within six months, which is about the time it will take you to build up a new business providing you work hard, it is not likely to be a worthwhile proposition.

One more consideration relating to goodwill is the publicity given to the business address and perhaps more importantly, in the case of a small one-car driving school, the telephone number. If the goodwill really only means the purchase of the existing clients at the time of buying the business and does not involve the transfer of the telephone number, then it is doubtful whether more than £500 should even be considered.

Where an existing school vehicle is included in the sale, the signs and dual controls should be valued at no more than around £50. The vehicle itself should be assessed separately and the mileage taken into account. Remember, the vehicle is only worth what you can get for it in part exchange for a new model.

Franchising

Franchising facilitates the marketing of a particular product or service at a quicker rate of expansion and on a larger scale than would otherwise be possible. It primarily cuts out the expensive middle management of large companies and, in effect, passes this responsibility on to the franchisee.

It is a method of starting up a business at less risk, whereby the franchisee buys the right to sell a tried and tested product or service of the franchising company and to carry on the business using the name and other services provided by or associated with the franchisor. The franchisor provides a business package which is operated within a specified area or location under the corporate name of the parent company.

Franchisees should be provided with training and expert backing which would not normally be available to the person setting up his own business, on precisely how the operation is to be run and on the standards to be maintained. In the case of a driving school, a franchisor should provide a complete 'blueprint' for every aspect of operating the business. This should include:

Complete training in instructional skills
Complete training in the business operation

backed up by:

Fully documented training programmes
Teaching notes
Student notes and other literature
Training aids for the learner
A business administration system
Management and business advice
Regular visits by experts from the franchising company
A system of supervising and monitoring standards
A system of communication between franchisees
Regional or national promotions by the franchisor
Regional and national advertising by the franchisor of the service.

A franchise normally contains two cost elements: a pre-launch service involving an initial payment for training, equipment, documentation and services to set up the business; a continuous service, after the initial launch, which normally involves commission charges on the services provided and/or payments for the products supplied by the franchisor and used in the day-to-day operation of the business.

The system of monitoring the standards of the individual franchisees may at first seem an unnecessary imposition. However, any franchise offered without such monitoring should be avoided. Monitoring is absolutely essential to protect the interest of each individual franchisee and also the reputation of the parent company.

Where adequate controls do not exist, a bad franchisee will have an adverse effect on the overall name and reputation of the services provided, which can result in a chain reaction on the trading of other franchisees. Once obtained, bad reputations are difficult to get rid of!

Avoid those franchisors with heavy 'enrolment' fees and those seeming to charge no fees at all. Where the ongoing commissions or fees seem too low to support the claims of the parent company, particularly in relation to its monitoring of the franchisees' operations, the company should be avoided. It is always wise to meet an existing franchisee before committing oneself. Avoid the company who will not make such an introduction for you.

Before making decisions either to franchise out a particular service which *you* have developed, or to become a franchisee it would be wise to obtain professional advice. *Taking up a Franchise*, published by Kogan Page, is recommended, and two booklets which may help can be obtained from Franchise World, 37 Nottingham Road, London SW17. They are 'How to Evaluate a Franchise' and 'How to Franchise Your Business'.

In practice, and with the exception of the British School of Motoring, franchisors do not have a good reputation within the driving school industry and have so far failed to provide franchisees with the expertise and services required and for which they have often paid. It has been estimated that the minimum cost of setting up a soundly-based franchising company within this field, with all the necessary services and training, would be in the region of £500,000.

Starting your own school

Starting up a driving school does not require a large capital outlay or involve vast amounts of administrative work. However, it does require a simple understanding of the principles involved in finance, driving school administration and the marketing of the services. If you would like to work for yourself as the proprietor of your own driving school then your priorities should be:

1. To qualify as a Department of Transport Approved Driving Instructor.
2. To devise your scheme for operating the business, including financial calculations, the courses you are going to offer, means of attracting business etc.

As most driving schools are started from scratch, most of the information in this book is slanted towards such an establishment.

Legal structure

There are three principal types of business. Most driving instructors operate within the first category.

1. The sole trader

There is very little to prevent anyone running a business within this category. It is not even necessary to register the business name. Care should be taken in choosing a name, however, and those which are likely to imply association with the Crown, the government, a local authority, or any which are likely to mislead, should be avoided.

Avoid choosing a name under which someone else in the area is operating. A quick look through the Yellow Pages, industrial and commercial directories, will help you in this respect.

It is advisable to choose a name which describes the business you are in such as 'Smith School of Motoring', 'Smith Driving School' or 'Smith—Driving Tuition' etc.

If you choose a name other than your own personal title, you must display the owner's name prominently on the business premises and on all business stationery. The notice on the premises should read: 'Particulars of ownership of (trading name) as required by Section 29 of the Companies Act 1981. Full names of proprietors: (insert names). Addresses within Great Britain at which documents can be served on the business (insert addresses).' You must also provide this information on request to any customer or supplier.

If you want to protect a trade name or trade mark and prevent others from using it, you must have it registered at the Trade Marks Registry, Patent Office, 25 Southampton Buildings, London WC2A 1AY.

Operating under this category is suitable for businesses where there is little financial risk. It enables you to employ staff and take all the profits. It also means you must pay all the taxes and bear any losses. Creditors can claim against your personal possessions, which include your home, if you run into difficulties.

2. Partnership

A partnership is an agreement between two or more persons to run a business together. Each partner is individually responsible in law for business debts incurred by the business as a whole and the agreement is based on mutual trust. Otherwise, it is in many respects similar to the sole trader category. Taxes are paid on the partnership profits as a whole rather than by the individuals.

It is advisable to have a solicitor draw up a partnership agreement.

3. Private limited company

A company has similar rights to an individual and is a separate entity in its own right. It costs around £100 to £200 to incorporate a new company or buy a ready-made company name 'off the shelf'.

A company requires a minimum of two shareholders, one of whom must be a director. It also requires the appointment of a secretary. The shareholders own the company and appoint the directors.

All the debts incurred by the company remain the company's debts and if it goes bankrupt the shareholders are not personally liable beyond the face value of their shares. In practice, though, the directors have probably had to put up personal guarantees to obtain company loans and obviously are responsible for these.

A limited company is obliged to prepare annual

accounts for the inspector of taxes, which must be audited by an independent auditor; the profit and loss account and balance sheet must be included in an annual return to the Registrar of Companies, as well as information about share ownership and changes in the company during the year.

The objectives and operating rules of a limited company are set out in its memorandum and articles of association; do make sure that these objectives are expressed in broader terms than you may require at present, so you are covered if you wish to expand the business or diversify into other activities later on.

Chapter 2
Financing the School

You will require some ready cash or personal assets to finance the launch of a new driving school and you can hardly expect anyone else to have confidence in your ideas and provide capital if you are not prepared to risk any of your own. Wherever possible, finance the new school from your own resources; because of the relatively small amount of personal investment involved (due to the wide range of vehicle financing options) this is a realistic possibility in most instances.

In addition to private savings, there are other readily realisable assets such as jewellery, antiques and collectables; you may be able to borrow up to 90 per cent of the surrender value on your life insurance policies; you may be able to use your house for obtaining a second mortgage, though this option can be expensive and should be avoided if possible. If you need to borrow, only obtain what you need and when you need it.

Sources of finance

Private loans
Loans from relatives and others are often executed through a bank in the form of a guaranteed overdraft or loan without the guarantor having actually to hand over any money. However, if the terms of the overdraft or loan are not fulfilled, the guarantor is liable for the amount.

If you borrow directly from relatives it is better for all concerned to have a proper agreement drawn up by a solicitor covering the precise terms of the loan.

Banks

Although banks are in the business of lending money, the manager has a responsibility to his employers and the company's shareholders to ensure as far as he can that you are a good risk. It is his duty to establish that:

The money/loan is secure.
You will have sufficient cash flow to run the business.
The interest on the loan can be repaid without difficulty.

He will want to know how you intend to generate business, what your track record and qualifications are, and how much of your own money you are investing in the business. He will also want to know the services you intend to provide and precisely how the cash is going to be used. You will have to tell him what you are going to live on while you are building the business up.

You need a business plan, first for your own purposes, to see if it is viable; second, with projected figures appended to show your bank manager or other potential source of funds that you have thought it out in detail and know what you are talking about. More detail of the costs involved are given in Chapters 3 and 4. It is no good making an appointment to see the bank manager with a view to securing a loan until you have clearly established, in your own mind and on paper, that the business proposition is sound.

Overdraft

The most common form of help the bank will provide for a small business is the overdraft. This provides the facility to overdraw on an account up to an agreed amount specified by the bank. Interest is paid on the amount overdrawn, usually at about 2 to 4 per cent over the base lending rate.

The overdraft is a convenient way of financing a business through the ups and downs of receipts and expenses, and helps through the more difficult periods, but because an overdraft facility can be withdrawn at short notice it is inadvisable to finance medium- or long-term requirements in this manner.

Bank loans
These are fixed-term loans, normally secured against business or personal assets. Interest rates may be either fixed or variable. These loans are: short term (one to three years), medium term (three to ten years) and long term (10 to 20 years).

Bank start-up schemes. Most High Street banks have their own start-up schemes for small business. Ask for details and see what best fits your needs.

Government Loan Guarantee Scheme
The government guarantees up to 80 per cent of a loan to a lending bank where it feels the intended business is basically a sound proposition and the borrower does not possess the necessary collateral. The limit for this type of loan is £75,000. The borrower pays a premium of 3 per cent to the LGS fund for the proportion which it covers, over and above the bank interest rate, so it is not 'cheap' money.

Enterprise allowance scheme for the unemployed
This allowance, which was announced in the 1983 Budget, may be of interest to unemployed persons with viable plans to start up their own business. If you have been unemployed continuously for 13 weeks and can show you have £1000 which you are prepared to invest in the new business, a grant of £40 a week for a whole year is available from the Manpower Services Commission and administered through Jobcentres.

The business plan

Calculating your cash requirements is crucial, both in setting up the business and in its day-to-day operation. If you are planning to open a new school you will need to determine how much cash is required for its initial launch in addition to the costs involved in its administration and daily operation. These will include the fixed costs or overheads and the variable costs which change according to the amount of work carried out.

Hard cash will be required for the day-to-day running of the business and this is quite different from the profitability of that business. Failure to understand the difference between *cash flow* and *profit* can be disastrous. *Cash flow* is the amount of cash required to finance the business over a period of time. *Profit* is the overall surplus of income over expenditure at the end of a given period, usually calculated over a year.

Cash flow is not normally a particular problem for the busy driving school as most clients will pay for their lessons as they take them or even in advance for courses. Driving schools carrying out a large percentage of their work for companies, government departments or educational establishments, however, will be paid in arrears, and the consequences of this to a new business with inadequate financial reserves to carry it comfortably through the period before the cheques come in should be considered. (Although a new driving school is unlikely to be in this position to such an extent where it causes problems, it is something which should be clearly understood.)

You will have to calculate:

The costs of setting up the business
The fixed costs or overheads
The variable costs relating to business carried out.

The setting-up costs

This is the minimum expenditure required to start trading. It involves the initial cost of equipment, services and property. If you do not possess the cash resources to finance the launch of the business without leaving yourself short of working capital until work starts coming in, there are ways in which expenses can be cut down considerably by leasing rather than buying the car and other equipment. Chapters 3 and 4 give more information on what is involved. Always use borrowed money for equipment, and reserve your own funds for working capital.

Account must be taken of the cost of such items and services as:

- A suitable, taxed and insured tuition car fitted with dual controls and advertising signs
- Telephone and answering facilities, which might include external extensions, answering machine, a commercial answering service, paging service or radiophone where office answering facilities do not exist
- Office furniture, stationery and typewriter
- Initial advertising campaign
- Any initial costs involved in forming a limited company or partnership if either method of business operation is chosen.

Most people are already in possession of some of the essential equipment, such as the telephone and a suitable vehicle. If it is intended to run the business from home then this keeps the start-up costs low.

Even where the car already owned may be unsuitable for driving tuition, its capital value can be realised against a more desirable model. In any event, there are other more convenient ways of financing the vehicle and reducing the start-up capital required.

Other less essential, but desirable, facilities and equipment will vary according to the type of services you intend providing and can often wait until the business becomes more established, such as:

Purchase or lease of property
Property decoration, repair, fitting etc
Office equipment — some form of duplicating or copying equipment is desirable both for producing your own PR and advertising material and for hand-outs to clients.
Stocks of books and *Highway Code* for sale.
If group training sessions are planned, teaching aids will be required. Overhead projector and screen are among the most useful, video is also helpful and relatively cheap but available software is fairly limited.

Fixed costs and overheads
These are costs which are incurred whether you are working or not and include interest which must be paid on loans and overdrafts.

Administrative costs
Running the business will involve general costs such as stationery, postage, advertising and fees for professional services (such as accountants) and subscriptions to professional organisations.

Equipment costs
Rentals for telephone and associated equipment such as answering machines, hire purchase or lease payments on vehicles etc. (A significant reduction in the setting-up costs of the business can be achieved if the vehicle is paid for, say, over a two-year period or leased.)

On purchased equipment such as the car, there is a depreciation factor—it will wear out and have to be replaced. Even if the vehicle is not used it will still depreciate in value from the time it is bought. Reserves must be built up to pay for a replacement vehicle when the time comes. Twenty-five per cent of the original price or £2000, whichever is the higher, is the maximum depreciation allowable for tax purposes in any one year.

Cost of premises
If you own or lease premises used for the business, you will have costs such as rent, rates, insurance, cleaning, decoration, repairs and other services such as electricity and gas. (If you are running the business from your own home it may be inadvisable to claim all of these costs as it could make any subsequent sale of the property subject to capital gains tax. Your accountant will advise you on this and as to what proportion you can safely claim.)

Salaries, pensions and insurance
You will need living expenses, which will be drawn from the business. If your spouse is working for you,

answering the telephone, booking lessons and keeping accounts etc, he/she should be paid for this.

You will have to pay National Insurance contributions and make provision for pension and accident or sickness schemes in addition to holidays.

Variable or running costs
These are the costs incurred as a direct result of working and giving driving lessons and include:

Petrol and oil
Servicing
Repairs and spare parts
Tyres.

Checklist for your business proposal

- Decide the minimum weekly income you need to meet your existing financial commitments such as mortgage, rates, National Insurance, gas and electricity, housekeeping and hire purchase payments.
- Decide what additional financial commitments will be incurred as a result of starting up the new business.
- What provision do you intend making to secure your finances in the event of ill-health?
- An accident can put your car off the road! Have you considered the possible costs of temporary car hire?
- What is the level and standard of the local competition? Can you offer something different?
- Have you checked with the local Planning Department how the law may affect you and your new business, eg, possible changes in the use of either your home or office? If you plan to run the business from home, are there any restrictive covenants in the deeds to your property which prohibit this use? (There is more information in Chapter 4.)
- Remember to inform the local tax inspector of your changed circumstances.

You will find it valuable to write out a business proposal which you can submit to possible lenders and which will clarify your own ideas. It should include the following:

- Name, position and address of proposed proprietor.
- Details of the proposed business venture and its location.
- Information about the proprietor(s); a curriculum vitae with evidence of your ability to run the proposed business.
- Premises. Full information, including present use, and whether planning permission is necessary for your business proposal.
- The market research you have done as to the viability of your business in the proposed area.
- Your financial requirements for the entire setting up, showing how much money is coming from which sources.
- Information on the services you propose to offer (1) during the first six months (2) during the first year (3) beyond that.
- What staff you plan to employ (part or full time); even if it's just your family helping out, put it all in.
- Budget for the periods of operation enumerated above.
- Cash flow forecast for budget periods.
- Details of personal and financial references.

Chapter 3

Financing and Choosing the Vehicle

The school car will be the most costly item of expenditure incurred in setting up your new business. There are only two ways in which this can be financed. You either purchase or rent it. If a business acquires a vehicle under a contract which provides for the business to have, now or eventually, the title to the vehicle, it will be regarded as a purchase. If the contract does not lead to, or offer, the eventual title of the vehicle, then the contract is a rental.

There are various forms of purchase or rental:

Purchase	*Rental*
Cash purchase: Own savings Overdraft Bank loan	**Lease options:** Open ended Close ended
Hire purchase: Hire purchase Lease purchase	**Contract hire:** Without maintenance With maintenance ·
On changing vehicle: Residual value to owner	**On changing vehicle:** Residual value to the rental company
Interest charges are allowable against tax.	Hire charges are allowable against tax.
25 per cent of purchase price or £2000, whichever is the lower, is also allowable. Provisions are contained in tax legislation whereby 100 per cent of the purchase price can be set against tax in the year of acquisition, where the vehicle is used entirely for business purposes in some trades such as taxis and driving schools. (Ask your accountant.)	VAT is reclaimable on contract hire charges and lease finance charges. (This is only applicable where the business is registered to charge VAT to its customers and does not apply to the majority of small driving schools.)

Purchasing the vehicle

Cash purchase

There are basically two ways in which you can go shopping for a car with the intention of paying cash for its outright purchase.

First, you can finance it directly from your own resources, eg, from your savings, or by trading in an existing car and making up the balance from your own funds.

Alternatively, you may be able to secure an overdraft or bank loan. An overdraft may not be suitable for long-term purchases such as your vehicle because, although this normally involves the lowest interest rates, it could be called in by the bank at any time (under normal circumstances, this is unusual unless there is an excessively high risk involved). It may be more advisable to secure a bank loan, where the bank simply deposits the required amount into your current account, leaving you free to select the car of your choice and sign a cheque for its purchase.

Cash purchase is normally the cheapest way of financing a vehicle. You will receive a tax allowance on 25 per cent of the car's value or £2000 per year, whichever is the lower figure. If you take advantage of a bank loan for the purchase, you will also get a tax allowance on the interest paid on the loan.

If you are using your own personal savings, it should be borne in mind that the money used to purchase the car could be otherwise earning interest if invested. The loss of interest should be taken into account in assessing the true costs of cash purchase. Further, the cash is not available for other uses within the business should it be needed.

Hire purchase

You will normally require a 20 per cent deposit and will be allowed tax relief on the interest involved in the car's purchase and also on 25 per cent of its value or £2000 per year, whichever is the lower, and there is no VAT charge on the rental payments.

This is a more expensive way of purchasing a vehicle than obtaining a bank loan and buying it outright because the interest rates are considerably higher. Some manufacturers, however, at times offer driving schools special interest rates on hire purchase agreements. This method of purchase has the advantage that the car itself provides the security and is flexible in that the contract may be taken out for any period and is capable of being terminated early (in which case there will be an interest penalty).

Lease purchase
This type of contract enables you to purchase the vehicle at the end of the lease period. It will normally require an advance payment of three months' rental.

The main advantage is that lease purchase enables the vehicle to be purchased at a low initial cost which may be of significant value for the person starting up his own business. The price paid for this advantage is higher monthly payments. There is also a difference in that tax is calculated on the payments made and not on the total purchase price of the vehicle as in hire purchase.

Interest costs will be greater than for hire purchase because more money is outstanding throughout the lease period.

Renting the vehicle

Before deciding to rent your vehicle either by leasing or contract hire (outlined below) you should discuss the matter with your accountant.

Leasing
Leasing is currently a popular method used by large companies for financing vehicles. It is a long-term hire agreement and you never actually own the vehicle. Tax relief is allowed on the lease payments.

One reason why leasing is so popular with large companies is that it frees their capital resources for other purposes and VAT on lease payments is recoverable. This, however, will not apply to the smaller businesses not registered for VAT.

Some agreements will leave a residual value on the vehicle which must be repaid to the leasing company at the end of the lease period. Where you are able to obtain more than this value on the sale of the vehicle you are permitted to keep the difference. If, on the other hand, the sale does not realise the residual value then the balance must be paid out of your own pocket.

Contract hire—long and short term

In the initial stages of starting up, this method of financing the school car might seem very attractive because of its initial low cost. However, under normal circumstances, it is not recommended if an alternative exists, because it is usually the most expensive method of financing a vehicle in the long term.

Contracts may include just the provision of the vehicle, or the car and also the dual controls, road tax, insurance, repairs and tyres. Prices may sometimes be deceptively low: there could be hidden extras. The VAT is often not included in the prices quoted— remember that this is a real cost to the business which is not registered.

Other than the initial cost, there is no real advantage in this method of financing the vehicle, although hirers claim it is easier to determine and stabilise vehicle costs for the duration of the agreement. In the long term, the costs of re-hiring on a new contract will be relatively higher and the claimed advantages are questionable.

The car cannot become the property of the hirer unless the contract is converted to a purchase agreement, and if considering this method of financing the vehicle, it should be clearly established at the outset whether or not there is any mileage charge incorporated in the contract.

Used vehicles

An important point worth some consideration, which can reduce the initial costs of start-up and depreciation, is to buy a good used vehicle with low mileage around a year old. Purchasing a used car can be risky and if you buy it privately rather than from a dealer there is often no guarantee or legal protection. However, if you are looking for a bargain you are unlikely to get one through a dealer who will want his share of the profits made from any sale.

Buying second-hand vehicles for driving school use will require a considerable knowledge of prices and the ability to assess condition and quality, eg, recognising defective or accident damaged or repaired vehicles. The inexperienced man dealing in trade circles is likely to come unstuck in two areas.

- The first is that he may pay more than the vehicle is worth or more than he intended. One way of finding out the going rates might be to obtain last month's issue of Glass's guide (the car dealer's full list of car values) from someone in the trade. They are not made available to members of the general public.
  Alternatively, the motorist's guide to car prices can be obtained from a newsagent. This should not be relied upon for total accuracy however, particularly when dealing with the trade.
- The second is the question of quality and, in the absence of specific guarantees regarding mileage and accident damage etc, there may be no legal redress particularly if buying privately or from the car auctions.

The risks involved in buying from car auctions are normally very high but perhaps not so great, for example, when purchasing low mileage cars under 12 months old through the British car auctions 'Union Jack Car Sales', or comparable schemes when guarantees are given. The vehicles involved often come direct from manufacturers or reputable main dealers with mileage and quality guaranteed.

The savings involved in buying from the auctions, particularly where inexperienced bidders are concerned, may simply not be worth any risk at all. On the other hand, if you have the time to attend car auctions to learn the procedures before buying, it may sometimes be possible to get a bargain. When bidding, look to see what price more or less identical cars are fetching. Place your maximum bid somewhere in the middle—*and don't bid over this!* The auctioneer will recognise you are not a trade buyer and it is his job to get the highest price possible for the car. He will therefore try to push your price up further than you intended going. If you let this happen there is no point buying at the auction at all. You might just as well buy from a reputable dealer and have all the benefits of this in addition to those provided under the law. Unless you are experienced in this field and if you intend buying a car in this manner, try to find someone with experience to accompany you. Dealers at auctions have been known to bid for their own cars simply to push up the price.

Another source of used vehicles is the Channel Islands where, if you just happen to be over there on holiday in the autumn, you may be able to pick up a low mileage vehicle in a territory with an overall speed limit of 40 mph. The cars are delivered by sea to Portsmouth or Weymouth where you pick them up and pay the car tax and transport costs. VAT must also be paid on the vehicles. Because of the costs of travel and the red tape involved, unless you are buying a small fleet of vehicles and are making a holiday of the trip, it is unlikely to be profitable.

Selling the vehicle

A safer way to reduce the costs of depreciation is to sell your old vehicle privately when renewing it. This normally requires capital resources to finance the new vehicle before selling the old one. A double bonus is gained here as, in addition to the amount raised from the sale of the old car, you can also command a discount on the new model.

Choice of car

Selecting the correct tool for the job is an important part of any trade. To the driving instructor, the car is the instrument of his trade and its selection is a personal and often subjective matter.

Although the author would hesitate to say that any car is totally unsuited for all purposes relating to training in the broad sense, it should be obvious that some are better suited than others for tuition purposes.

Cost of the purchase will be a main consideration; another will be which models are likely to be superseded in the near future. When it comes to changing the car, the lower priced model may prove to be more costly to renew when depreciation and finding the difference between the old car and the replacement price of the new model are taken into account. In addition, the running costs may prove to be higher. The concept that 'small is beautiful' mostly holds good from a running costs angle, since servicing, repairs and replacement of such items as tyres, brakes and clutch will usually be lower. Insurance premiums too are likely to be lower on the smaller engine capacity models.

When purchasing the car, consideration must be given to the use to which it will be put. Will the small car withstand the rigours of the job? It is by no means certain that the smallest capacity engine will be the most fuel-efficient when put to the use of driving tuition.

Servicing costs and probably petrol consumption are more readily determinable. Residual value (the price you receive for the car in part exchange at the point of sale) is more difficult to assess and careful consideration should be given to this before making a choice. After fuel, this is likely to be the highest cost to the driving instructor.

The perfect driving school car probably does not exist and it is not the intention of this book to compare the faults and virtues of different makes, but some of the factors which might influence choice should be mentioned, such as:

Ease of driving
All-round visibility
Economy and reliability
Servicing arrangements and repairs
Comfort
Size and design.

1. Ease of driving

A good driving school car will have well-positioned

39

controls, central floor gear change with light positive gear positions and a well-protected reverse gear (to avoid accidental selection); a centrally mounted hand-brake is also desirable for ease of reach by the instructor in an emergency. The steering should be light and precise, the accelerator, footbrake and clutch smooth and progressive. Servo-assisted brakes are desirable and the engine should be smooth in both first and second gears at tick-over speed on a level road. It should also be comfortable in fourth at about 25 to 30 mph.

2. All-round visibility
The all-round visibility car will have a good sized interior mirror with two externally mounted door mirrors. In general terms, the higher the seating position, then the better the all-round vision and general feeling of security.

3. Economy and reliability
Economy is not just the price you pay for a car, you must also consider what price you might get when changing it. A cheap but unreliable car may turn out to be very expensive through lost lessons and repairs. Good mileage per gallon is of obvious importance and so is the cost of spares. When considering the cost of spares, it is also significant that it is less expensive to replace a clutch which has given you 50,000 miles of usage, than to pay half as much four times as often.

Find out the cost of servicing and check the frequency which the manufacturer recommends for this to be carried out. Some companies offer special incentives for instructors to buy their particular models, so make enquiries about this—but take a hard look at the offer's worth: you may be better with a cash discount. Check the insurance grouping and costs before committing yourself to any purchase, and try to establish what trade discounts the company is prepared to give on spares and servicing. *If you do this before you buy the car, you are in a stronger bargaining position.*

4. Servicing and repairs
If you intend servicing the vehicle yourself and carry-

ing out other minor maintenance work, establish beforehand how this may affect the manufacturer's warranty.

The availability of spares is extremely critical to the driving instructor, who cannot afford to have his car off the road because of some elusive part.

From the convenient servicing point of view, it is better for you to buy a vehicle from a local dealer providing you are able to come to mutually agreeable terms. You will need instant service when anything goes wrong. Of course, everyone wants instant service when his car breaks down, but a tuition vehicle off the road unnecessarily waiting its turn means lost income for each hour of the day the car is idle. Make sure, before buying the vehicle, that you are going to receive preferential treatment. If this is not agreed upon, you may do better to buy a little further afield.

5. Comfort
This is extremely important when choosing a tuition car. Remember, you may sit in it for several hours at a time. From the driving point of view, ensure that the seat provides adequate support and has a wide range of adjustment to suit a variety of shapes and sizes. People with small feet may find high pedal positions tiring and troublesome; a noisy car creates stress and will tend to have the overall effect of leaving the instructor more tired at the end of the day. Short-term financial gain at the expense of instructor comfort is false economy; a tired instructor with backache is seldom capable of giving his or her best.

6. Size and design
People often choose the wrong car because they like the look of it. We are all human and the manufacturers are well aware of this. It is well known also that instructors often buy the car of their personal choice when it is not really suited to driving tuition.

It is very difficult to know quite how to offer advice in this respect because of its subjective nature and also because of the strong motivational factors influencing the instructor's decision. However, as to the best size for a tuition car, the small to medium models are most popular, eg, from the Mini up to the Volvo 343. The large cars, in the author's opinion, tend to be rather unwieldy for tuition and the very small ones too noisy and not altogether comfortable.

Clients will themselves have their own opinions and preferences but, generally speaking, they tend to feel more secure in a sturdy, medium-sized car which provides them with good all-round vision. Obviously in these days of economising, the fuel consumption of smaller cars is a serious consideration.

Alternative fuels

Gas conversions are becoming quite popular with driving instructors and significant savings in fuel costs can be achieved. The cost of conversion can be comfortably recovered in less than a year.

Diesel engines are far more efficient than the petrol engine and considerable savings can be made in fuel costs. However, diesel-engined vehicles cost more to buy and are somewhat sluggish in performance. While some driving schools do use such vehicles, they are not particularly popular as there are characteristics associated with the accelerator which make it arguably less suited to the new driver.

Dual controls and instructor's mirrors

There are two main types of dual controls, namely cable and rod operated. Most instructors seem to prefer the rod type because they supposedly provide more 'feel', but with the modern cable controls this claim is questionable.

Rod controls are generally more expensive and more difficult to fit. They also have to be changed more frequently as even slight differences in car design can make them obsolete. These controls tend sometimes to interfere with the driver's pedal operation due to poor location of the operating arms. On some, it is also possible for the instructor to get his feet trapped under the dual pedals when the car controls are being operated by the driver.

Cable controls are more versatile in this respect and can normally be adapted to most models without new sets being required. It is recommended, however, that new cables are fitted whenever the controls are changed from one vehicle to another.

Rear view mirrors and those for checking the pupils' eye movements are available from most accessory shops. The John Sydney 'Trafalgar' brand are good, sturdy, vibration-free mirrors. *Note:* they should be located so as not to impede the driver's view.

Chapter 4

The Driving School Office and How to Run it

Unless you have large capital reserves to back you up, it would be prudent to start up from your own home and establish a full book of clients before taking on the added expense of rent, rates, reception and other costs involved in running an office at a time when capital is going to be short.

There is also the possibility that you may be able to share the costs of an existing office with, say, an insurance broker who wants to attract new business into his premises, or even a hairdressing salon may sometimes be appropriate where space and facilities permit. However, just a card on the counter or reception desk of an existing business is a non-starter and will not create a very good professional image.

Office location and premises

Almost without exception, if you want to sustain sufficient business to operate three or more cars consistently, you will require a centrally located office or shop window or one in a busy area and clearly noticeable to passing traffic and pedestrians. Ideally, the office should be on a corner or facing a road junction. Facing a busy T junction is a particularly good spot in which to get noticed quickly. The office which is tucked away on a local housing estate is unlikely to do well.

The appearance of the office should be very presentable. Shabby and badly decorated driving school offices give the impression that the service provided will also be shabby. The same will also apply to the condition of the driving school's cars.

Coloured perspex or similar material for fascia boards can be obtained from local stockists and used to smarten up an otherwise fairly old and dilapidated property. Very effective signs can be made up for you if you have the skill to erect them yourself.

There are various other ways by which you can cut the cost of expensive shopfitting services, which do provide the most professional image but may generally be unnecessary if you are good with your hands or have some contacts who might be able to do the work for you in their spare time.

The office window should look professional and attractive in order to catch the attention of passers-by and make them stop to look in. Remember, everyone who stops may be a potential customer or an advertiser for your services! Posters, models, certificates, qualifications, photographs of successful clients are all things which interest passers-by. Automatic rear projection screens showing road scenes and driving themes can be used to attract a tremendous amount of attention at very low cost if you have this equipment, which can also double for slide projection equipment in your classroom.

Other considerations should be taken into account such as sanitation amenities for staff and pupils, particularly if classroom or group training facilities are planned. Obviously, size is a factor which will be reflected in the rent, rates and running costs. If the provision of reception facilities only is envisaged then a small area will suffice. This will help to keep costs to a minimum. If you do intend providing classroom or group training facilities then a reasonably large space will be required.

An area of around 14 by 20 feet will be quite adequate for most purposes where groups of around 10 or 12 are involved. Groups over this size are not recommended under normal circumstances particularly for instructors who are not experienced in a classroom environment. Large rooms or halls are not really suitable for this use as they cost too much to heat and tend to provide a very formal atmosphere which may frighten students away. It is doubtful whether an area less than 12 by 18 feet will be totally suited for such purposes when projection lengths etc and seating capacities are taken into account.

Before embarking upon any scheme to provide classroom facilities two important factors must be taken into account. First, there is a public resistance to classroom tuition, although it can be made to work. Second, teaching a group of students in a classroom or workshop environment is more complex than most people imagine and involves quite different skills from teaching in a one-to-one environment in the car. If you have these skills it is a good way of attracting business which might otherwise go to your competitors. If you do not have them, it might be better to leave such plans until such time that they have been acquired.

There are tremendous opportunities to present a more professional image than your existing competitors. It's really a matter of whether you are willing to accept the challenge and have sufficient vigour, stamina and enthusiasm to carry it through.

Statutory requirements affecting the use of premises for driving instruction

All 'development' requires planning permission under the Town and Country Planning Act 1971. For a definition of 'development' and an interpretation of what constitutes it, the local Planning Department should be consulted or professional advice may be needed. In principle, the change of use of a building or land or the erection of a building requires planning permission. There are exceptions to this principle in the following circumstances:

1. Under the General Development Order 1981 deemed permission is given for certain minor operations which are defined in the Order. These are not likely to affect premises used for driving instruction purposes, except that they do allow certain extensions to dwelling houses, which may be relevant in respect of 2. below.

2. It does not constitute development for the *use* of any *building* or *land* within the boundary of a dwelling house for any purpose incidental to the enjoyment of the dwelling house as a dwelling house. For example, a professional man (if

his deeds allow him) can use his premises in pursuit of his profession, provided he does not employ people other than his family, eg, a doctor can see patients at his house, a writer may write books, a school teacher mark books etc. This principle is long established and a driving instructor operating from his own home, not employing others, is not likely to be challenged by a local Planning Department as to the need for planning permission.

3. In respect of other premises, then the use of a building for driving instruction purposes would constitute 'a change of use' and require permission unless it were an existing use, ie, an 'office' as defined by the Town and Country Planning (Use Classes) Order 1972, which puts driving instruction in the 'office' of class II. *Note:* 'Office' in this context does not include a post office or betting office, therefore such existing premises would require planning permission to be used for driving instruction purposes. However, any other building in class II of the General Development Order would not need planning permission.

The right to use premises for driving instruction bestowed under the above deemed permission of the Act does not mean that physical external alterations or external signs would not need permission: if they materially (in the planning sense) affect the appearance of the building, they would. For example, premises previously used as a bank, estate agency, building society etc are class II uses, therefore to use them for driving instruction would not constitute a change of use requiring permission, but to put a new shop front in *would* require planning permission.

How to apply for permission
From the above it is clear that, before any purchase or lease commitment is undertaken, the planning situation must be established. A visit to the local Planning Department will usually establish whether permission is required or not, but independent professional

advice may well be worthwhile, especially in doubtful circumstances.

Assuming planning permission is required other than for a change of use only, ie, no other external changes are to be made, then a form TCP1 obtained from the local Planning Department should be completed and a map to a scale of 1:1250 (which can usually be obtained by a private individual from the local authority) identifying the site by a red line around the plot and the appropriate fee (currently £44) should be sent to the local Planning Department. They are then normally required to give their decision within two calendar months of receipt of the application. If they fail to do so, then the applicant can assume a refusal and exercise his right to appeal to the Secretary of State against this non-determination. If he is refused permission, or given permission with conditions he does not accept, he can also appeal against the decision to the Secretary of State. No fee is charged for appeals.

If use of the premises requires material external alterations, whether it has existing use rights or not, then the application and appeal process are exactly as above but details of the changes to a scale of 1:100 have to be supplied additionally.

There are exceptions to the above general rights and procedures in respect of 'listed' buildings or buildings within 'conservation areas'.

It should be remembered that all changes of use and physical alterations, whether they require planning permission or not, do require permission under the Building Regulations of the Public Health Act. Your architect or builder will obtain these for you, and he must be instructed to do this before the work is commenced.

Should your premises require planning permission, then the considerations that the Planning Department are likely to take into account are as follows:

1. Will it cause nuisance to adjoining owners from a planning point of view?
2. Are highway problems likely to arise, eg, will it generate car parking problems (both driving

school staff and pupils)? Will picking up and dropping pupils cause traffic hazards or congestion, particularly in the rush hour tidal traffic flows?

Whether the school operates a pick-up service or not would not affect the local Planning Department taking this view, because once permission is granted it applies to the land/buildings and not to an applicant's current practice. This is because another operator may have a different policy but enjoy the planning permission.

3. Will associated classroom teaching generate traffic and parking problems with numbers of pupils attending a session?
4. Will the proposed building changes materially affect the visual character of the area?

These issues should be taken into account when seeking premises, for the satisfactory resolution of them before pursuing the other issues may save time, delays or abortive professional fees in trying to convince the local Planning Department that the premises are suitable.

Office administration

A driving school does not normally involve a tremendous amount of administration and bookwork and certainly, at the outset, these will be fairly simple and straightforward. The primary task is booking the lessons.

Lesson booking system and accounting
One of the first priorities for the self-employed driving instructor is an efficient lesson booking system to ensure he gets to the right client at the correct time and also that the client is there at the proper time. Students should be provided with a professional appointment card reminding them of the date, time and place etc of the lessons. The different service options are discussed on page 90.

Bad timekeeping and unkept appointments by the

instructor are guaranteed to lose customers, and appointments forgotten by the client may well mean financial loss to the instructor unless this is covered within the conditions of the service and the client is provided with a proper written reminder of the arrangements.

A good driving school will also keep records of how each pupil is progressing. These should show the strengths and weaknesses, what has been covered so far on the course and the routes used. This is the only way an instructor can possibly remember so much detail where he may have 40 or 50 clients, all at different stages of learning.

Independent records should be kept in respect of each client, showing advance payments, ordinary payments and lessons received. Where lessons are conducted on a credit system (usually only to government departments, eg, skill centres, Department of Health and Social Security, gas and electricity boards, education departments etc) separate records will be required together with an efficient invoicing/ billing system. You will need to keep a sales ledger. Each customer has a separate page headed with the name, address, telephone number and contact's name so they are readily available when you need to send a statement or chase payment by phone. Credit sales are entered on the left-hand side of the ledger, and payments on the right.

It is also advisable to keep records of daily receipts in respect of each instructor employed and lessons given. These can be conveniently kept with the relevant vehicle records. The system should be designed to avoid temptation for the employee and to provide an adequate record of receipts and costs.

Serious debts are not normally a problem in the driving school industry because the nature of payment does not usually involve credit. Facilities do exist, however, to obtain payment of debts up to £500 in the small claims section of the County Court. Any claims under £500 are normally referred to arbitration and need not involve solicitor's costs. The court can deal with claims arising from:

Non-payment for goods or services
The provision of deficient goods or services
Negligence, such as those resulting from a motor accident, loss of no claims bonus etc
Non-payment of wages or salary owing or payable in lieu of notice etc.

Office services and records will involve:

Receptionist/telephone answering service
Main office diary
In-car diary (for each car)
Student appointment and record cards
Student progress/record/assessment forms
Driving test booking records.

'Off the shelf' systems are available from the Driving Instructors' Association.

Telephone
The cost of installing a business phone is currently £85 plus VAT compared with £80 for a domestic system. The quarterly rate for the business phone is £22 plus VAT compared with the £14.15 plus VAT for the domestic. The business phone is strongly advised despite the extra cost because it ensures quicker installation and it covers a business entry in both the ordinary directory and the Yellow Pages.

Telephone answering machine
This is a poor substitute for the person who can both answer questions and book lessons. However, where a phone is left unattended for long periods, particularly during office hours, some alternative has to be found if business is not to be lost to competitors. The answering machine is better than no answer at all, because the machine can ask the caller to leave a message or give a time for calling back. The main problem with them is that people do not generally like talking to machines and more often than not they simply hang up. A properly delivered message, however, on a good tone quality machine can increase the number of people prepared to talk.

Many machines on the market are illegal and the

Sunday	Monday Date	Tuesday Date	Wednesday Date	Thursday Date	Friday Date	Saturday Date
	8	8	8	8	8	8
	9	9	9	9	9	9
	10	10	10	10	10	10
	11	11	11	11	11	11
	12	12	12	12	12	12
	1	1	1	1	1	1
	2	2	2	2	2	2
	3	3	3	3	3	3
	4	4	4	4	4	4
	5.30	5.30	5.30	5.30	5.30	5.30
	6.30	6.30	6.30	6.30	6.30	6.30
	7.30	7.30	7.30	7.30	7.30	7.30
	8.30	8.30	8.30	8.30	8.30	8.30

Name	Name	Name
Address	Address	Address
Tel. No.	Tel. No.	Tel. No.

Lesson booking diary, loose leaf system
(Courtesy Autodriva Training Systems)

FAULT MARKING CODE

- / Minor
- X Potentially Serious
- O Serious

APPOINTMENTS						GENERAL			CONTROL SKILLS				CONTROL OF SPEED				CONTROL PROCEDURES						OBSERVATN'N				CONSIDERATION							POSITIONING							Rd. JUNCTIONS CROSS ROADS					SPECIAL RULES					EXCER-CISES		
Day	Date	Time	£	p	Sig.	Application Highway Code	Steering Method	Starting Precautions	Moving Off	Stopping-Control/Accuracy	Parking Positions	Emergency Stop	Accelerator and Clutch	Accelerator and Gears	Brake and Gears	Handbrake Operation	Restraint in use of Speed	Precaution-Potential Hazard	Hangback Drill	Low Speed Control	Securing the Vehicle	Undue Hesitation	Mirror	Signals	Traffic Signs	Road Markings	Safety of Pedestrians	Pedestrian Crossings	Safety Clearances	Meeting Oncoming Traffic	Overtake Safely	Anticipation	Following Distance	Normal Driving Position	Lane Selection	Lane Discipline	Erratic Steering	Position-Turning Left	Position-Turning Right	Cutting Corners	Observations Before Emerging	Emerge with due Regard	Give Way to Oncoming Traffic	Precendence to Pedestrians	Precautions-Minor Crossroads	Contested Right Turn	Traffic Lights	Roundabouts	Dual Carriageways	One Way Streets	3 Point Turn	Reversing	Reverse Parking

Lesson appointment and student progress record
(Courtesy Autodriva Training Systems)

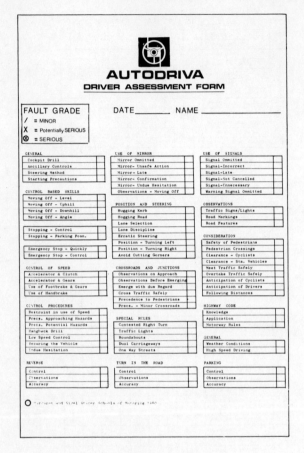

Driver assessment form

PROGRESS REPORT																						
CONTROL												ROAD PROCEDURE										
1 2 3 Date			4						5	6	7	8	9	10	11	12	13	14				15 16 17 18 19 20
			A	C	F	G	H	S										i	ii	iii	iv	

1. Comply with the requirements of the eyesight test.
2. Know the Highway Code.
3. Take proper precautions before starting the engine.
4. Make proper use of/accelerator/ clutch/gears/footbrake/handbrake/ steering.
5. Move away/safely/under control.
6. Stop the vehicle in an emergency/ promptly/under control/making proper use of front brake.
7. Reverse into a limited opening either to the right or left/under control/with due regard for other road users.
8. Turn round by means of forward and reverse gears/under control/ with due regard for other road users.
9. Make effective use of mirror(s) well before; take effective rear observation well before: signalling/changing direction/ slowing down or stopping.
10. Give signals/where necessary/ correctly/in good time.
11. Take prompt and appropriate action on all/traffic signs/road markings/traffic lights/signals given by traffic controllers/other road users.

12. Exercise proper care in the use of speed.
13. Make progress by/driving at a speed appropriate to the road and traffic conditions/avoiding undue hesitancy.
14. Act properly at road junctions:
   (i) regulate speed correctly on approach;
   (ii) take effective observation before emerging;
   (iii) position the vehicle correctly/before turning right/before turning left;
   (iv) avoid cutting right hand corners.
15. Overtake/meet/cross the path of/other vehicles safely.
16. Position the vehicle correctly during normal driving.
17. Allow adequate clearance to stationary vehicles.
18. Take appropriate action at pedestrian crossings.
19. Select a safe position for normal stops.
20. Show awareness and anticipation of the actions of/pedestrians/cyclists/drivers.

Student progress record
(Courtesy Autodriva Training Systems)

tone quality is not good. A British Telecom machine costs £18 for installation and £30 rental per quarter. BT licensed machines are also available from other sources—look in the Yellow Pages.

Telephone answering service
If you have a friend, neighbour or relative living nearby who is at home most of the day, you may be able to get British Telecom to fit a special system which enables your phone to be switched through to theirs while you are out, or even a simple extension can be fitted if they only live next door.

Commercial answering services are available in most areas. These are companies or individuals who will take your calls which you can either collect later or have them phoned through to you at a pre-arranged time.

Radiophones and paging services
National and international telephone calls can now be made from your car without going through the operator. The installations are compact and comprise a control unit, handset and a microprocessor/radio transmitter/receiver which is concealed in the car boot. Incoming calls are received automatically from any telephone in the UK. A wide choice of financial options is available, which includes leasing, lease purchase and rental.

Other types of in-car phone systems are used by some driving schools to solve their problems. One of these may be the answer to yours! However they are expensive, costing up to around £1000 a year.

The radio paging service is a cross between the radiophone and the telephone answering system. When you are paged you know that there is a call or message waiting for you.

Office typewriter and stationery
When starting up a new business, you need to become known by as many people as possible. You will want to create a proper professional image for the new school and write to local companies, youth clubs, schools, social clubs, newspapers and others. Circu-

late your friends and relatives and ask them to give out your business cards to people they know who want driving lessons.

To do all of this you will need properly designed and printed letterheads and business cards. You will also need a typewriter and preferably some way of quickly duplicating material for distribution.

You will require appointment and record cards which you can buy ready printed from a supplier if you don't want the added expense of originating your own.

Two lesson diaries are necessary (one by the office phone and one for the car), and a double entry cash book in which to keep your accounts.

By shopping carefully around and buying second-hand, you should be able to pick up a typewriter and duplicator for less than £100 (if you do not yet want the expense: a dry photocopier). The stationery need cost no more than around £50.

If you have your 'house style' logo designed professionally it can be expensive but it will prove worthwhile in the long term.

Vehicle administration

As the owner or registered keeper of additional vehicles, you have a legal responsibility to ensure they are serviced and maintained in a roadworthy condition, whether you are driving them or not.

A system of recording mileage, fuel used, servicing and repairs carried out, together with lessons given, is desirable. Where vehicles are being used by employees it would also be wise to keep a record of their private mileage separate from lesson mileage.

Breakdown and accident procedures
A clearly defined breakdown and accident procedure should be known by all staff. An outline for such procedures might be:

Breakdowns. If a vehicle becomes unroadworthy during a lesson, eg, a bulb becomes fused, the fault must

57

be rectified immediately before continuing. (Spare bulbs should always be carried.)

If possible, in the event of any type of breakdown, the vehicle should be brought to a stop at a safe place where it will not inconvenience others. For more serious delays, and where possible, the office should be informed so that the next client can be contacted.

Accident procedure. If an accident occurs during a driving lesson involving a vehicle belonging to the business, regardless of blame the pupil should be reassured. Steps should be taken to warn others of the hazard and warning lights and triangles, where available, should be used. Lights should not be obstructed, particularly in poor or dark conditions. The usual particulars should be exchanged with the other driver and any witnesses' names should be obtained.

If the vehicle is immobile, arrangements should be made immediately to have it taken away and it would be wise to remove any advertising signs etc. The office should be informed right away so that no other client is kept waiting, and arrangements made for the conveyance of the pupil involved in the accident either to home or office.

Vehicle running costs

Costs are incurred with the tuition vehicle, whether it is operating or not. Remember, you must make provision for its replacement when it is no longer economical to keep it running.

Where you have purchased the vehicle, consider:

Depreciation, which is the difference between what you pay for a new car in 1984 and its value when you sell it in, say, 1986; eg, a car costing £5000 today may realise less than £2000 in two years' time, and

The replacement car you buy in two years' time may cost £6000 for a similar model. You will see that the total difference to find is now £4000.

Fuel

Fuel consumption should be estimated at approximately one-half gallon per lesson. This will obviously

Daily Work Sheet

Date _____ Reg No _____ Instructor _____

Odometer reading day end _____

Odometer reading day start _____

Total mileage _____

Time	Student	Mileage covered	Route No	Fuel and Oil	Signature

Vehicle defect report

vary according to the size and efficiency of the vehicle's engine and the type of service which you intend to provide.

- *Operating from a central office.* Tuition cars cover far less mileage when operating in this manner where clients attend the office. In some instances, it may be possible to achieve three lessons to one gallon of fuel.
- *Overlapping pick-up service.* This type of service provides the students with maximum driving time per lesson but this is of questionable value to you as much of the lesson may be spent dropping off the previous client and picking up the next. The distances travelled will be considerably greater, thus increasing fuel consumption. Careful planning is required when booking clients in relation to the distances between.
- *Personalised pick-up service.* This type of service means that the client may lose the time it takes the instructor to reach his pick-up point. Providing the lessons have been organised in a sensible manner the time loss is generally insignificant and is usually more than compensated for by positive and constructive tuition throughout the whole period. This type of service does, however, increase fuel consumption but generally lends itself to a more professional service than the overlapping pick-up. Generally, two lessons to one gallon of fuel can be expected.

Tax and insurance
The current cost of a vehicle excise licence is £90 per year. Vehicle insurance costs vary from around £100 per year with maximum no claims bonus and also according to vehicle group and area of operation.

Insurance for tuition vehicles *must* include cover for the additional business risks. Normal comprehensive cover is inadequate and it is advisable to obtain a quote from a specialist broker dealing with driving instructor policies, such as the DIA (telephone 0483 65124).

Summary of Insurance Cover

Insurance cover available	Type of policy and limit of cover			
	ACT ONLY	THIRD PARTY ONLY	THIRD PARTY, FIRE AND THEFT	COMPREHENSIVE
1. Liability for injuries to other people				
2. Liability for injuries to passengers in your car				
3. Liability of passengers in your car for accidents caused by them				
4. Liability for damage to other people's property, including vehicles				
5. Liability for injuries to other people, or damage to their property caused by a trailer attached to your vehicle				
6. Fire or theft of the vehicle				
7. Accidental damage to the car				
8. Accidents to yourself, and possibly your wife or husband – may apply to all journeys in your car or any other car				
9. Medical expenses incurred as a result of an accident involving your car				
10. Loss of, or damage to, clothing and other personal effects (up to a stated amount) while they are in the car				

Courtesy of British Insurance Association

Car signs
Properly lettered, professional vehicle signs should be considered a must for anyone starting or building up a new school. The aim is to project a good professional image and get yourself seen and recognised around town as quickly as possible. Signs are available in various designs and from a number of suppliers, and include roof rack type, magnetic roof type, illuminated signs, Department of Transport ADI side emblems, magnetic side panels and various types of 'L' plates and emblems.

Costs are around £50 for magnetic signs and side panels.

Accounting systems

When operating a business on your own account you should always know the current state of your finances, what your running expenses are and what they are likely to be in the future. Accounting systems are devised to make this information readily available.

Accounts must be presented annually to the Inland Revenue for income tax assessments. If they are not, the tax inspector will make an estimated assessment which will probably be far higher. Books can either be presented direct to your local taxation officer or the services of a chartered accountant can be obtained. It is advisable to use an accountant who will advise you what can be offset against tax and what deductible expenses are allowed.

Accountants' fees will be kept to a minimum where books are maintained in an orderly manner with all income and expenditure itemised in balanced lists.

The simplest book-keeping is by double-entry method. Obtain a Simplex cash book and enter (usually on a weekly basis) all income (whether by cash or cheque) on the left-hand page and expenditure on the right, with a banking column for each. Check this monthly against your bank statement to ensure you have the basic information right.

Keep receipts in date order and enter them into the cash book regularly. Items such as postage, car wash, car parking etc for which receipts are not usually given, should be noted on petty cash vouchers.

The following is a list of income and expenditure likely to be incurred over a week in the running of a single-car driving school. The table following the list shows how these items should be entered into the cash book described above.

Cash book

Date	Item		Amount £
1 Jan	(1)	Cash in hand brought forward from 31 December	10.00
1	(2)	Balance at bank brought forward from 31 December	1,225.00
2	(3)	Petrol (cash)	10.00
2	(4)	Driving test fees (cash received from clients)	26.00
3	(5)	Test fees paid to Department of Transport on behalf of clients (cheque)	26.00
4	(6)	Purchase of spare parts for car (cheque)	15.00
5	(7)	Purchase of car cleaning materials (cash)	5.00
5	(8)	Petrol (cash)	15.00
6	(9)	Wife's wages (for answering telephone, organising lessons, paperwork etc) (cash)	25.00
6	(10)	Own wages (cheque paid into private account)	120.00
7	(11)	Telephone bill (cheque)	65.00
7	(12)	Local weekly newspaper advertising account for December (cheque)	40.00
7	(13)	Income from clients for the week (cash and cheques)	300.00
7	(14)	Cheque received from county council for running evening class on driver education	100.00
7	(15)	Local garage—petrol account for December (cheque)	110.00
7	(16)	Cash (and cheques) transferred to bank	370.00
7	(17)	Postage and sundries for week	1.00

Receipts

Date	Item		Cash £	Bank £
1 Jan	(1)	Cash in hand (surplus cash brought forward from 31 December)	10.00	
1	(2)	Balance at bank (money left in bank at the end of the previous week's trading)		1,225.00
2	(4)	Driving test fees received in cash from clients	26.00	
7	(13)	Income received from clients for week (cash and cheques)	300.00	
7	(14)	Fee received from county council for running course (cheque)	100.00	
7	(17)	Cash transfer to bank (cash and cheque paid into bank from income)		370.00
		Balance	436.00	1,595.00

Payments

Date	Item		Cash £	Bank £
2 Jan	(3)	Purchase of petrol (cash)	10.00	
2	(5)	Test fees paid to the Department of Transport on behalf of clients (cheque)		26.00
4	(6)	Purchase of spare parts for car (cheque)		15.00
5	(7)	Purchase of car cleaning materials (cash)	5.00	
5	(8)	Purchase of petrol (cash)	15.00	
6	(9)	Wife's wages (cash)	25.00	
6	(10)	Own wages (cheque paid into private account)		120.00
7	(11)	Quarterly telephone bill (cheque)		65.00
7	(12)	Local weekly newspaper advertising account for December (cheque)		40.00

Payments

Date	Item		Cash £	Bank £
7	(15)	Local garage—December petrol account (cheque)		110.00
7	(16)	Cash transfer to bank	370.00	
7	(18)	Postage and sundries for week	1.00	
			426.00	376.00
7		Cash in hand and at bank to carry forward to 8 January	10.00	1,219.00
		Balance	436.00	1,595.00

The balance may not show a true picture of the actual amount, since direct debits (car payments etc, paid directly from your bank account) are not generally entered into weekly/monthly accounts. These payments must therefore be considered when working out the actual trading status of the business. It is also useful to make a list of these payments for the accountant when the books are presented for audit as this will save his time and therefore your money!

If you wish to see at a glance the position of your bank account, a weekly record can be kept as follows. (This table relates to the foregoing Simplex cash book figures.)

Bank record week ending 7 January

			£
Item (2)	Balance brought forward from 31 December (credit)		1,225.00
Item (17)	Weekly cash and cheque transfer to bank (credit)		370.00
		Total	1,595.00

	£	
Cheques paid during week (debit)	376.00	
Bank payments (direct debit for car)	200.00	
Cash drawn from bank	nil	
Total		576.00
Balance at bank to carry forward to 8 January		£1,019.00

You will see from this example that there is a difference of £200 in the actual bank balance and the balance shown in the ledger — this takes into account the direct debit payment for the car.

Monthly balances will vary considerably since bills for the telephone, heating, lighting, office rent and rates etc, are usually paid quarterly. Allowances must therefore be made for all of these contingencies. January and July can sometimes incur more expense since income tax usually becomes due for payment at these times.

Also to be taken into consideration are the dates when insurance and road tax on school vehicles become renewable as these are expensive items and should be allowed for in the business budget.

Taxes and National Insurance

Value added tax
If annual turnover exceeds £18,700 then you will have to be registered for VAT, and your accounts will have to show all payments and claims relating to it. VAT is administered by HM Customs and Excise, who will provide you with the necessary paperwork, instruction leaflet, and show you how to deal with it. VAT returns usually have to be completed quarterly for inspection.

Very few driving schools are registered for VAT, probably because it puts them at an immediate price disadvantage over their competitors and involves more paperwork which they prefer not to do. This means, though, that they are unable to claim back VAT on petrol and other goods and services. VAT on car purchase cannot be claimed back in any event.

Most small schools deliberately keep under the £18,700 annual ceiling. Husband and wife teams will often run two separate schools simply to ensure the VAT ceiling is not exceeded as it would almost certainly be if they ran a joint business. It is not difficult to remain under the limits when running just one car and yet still make a comfortable living. On a 40-hour week (say 2000 lessons per year), an established business is unlikely to have expenses in excess of £5000 to £6000 per year leaving much of the remainder to live off. An added bonus is that the vehicle costs, which normally take up a considerable amount of the average worker's pay, are mostly reclaimable as a business expense. This represents a considerable saving on the demands made from personal earnings.

Income Tax

When you change from being an employee to a self-employed person, the manner in which tax is paid will also change. The local tax officer must be informed of the changed circumstances. Tax will be paid under Schedule D, which provides a wider range of business expenses allowable against tax.

The main allowable expenses applicable to the self-employed driving instructor will include:

- Running costs, such as telephone, advertising, stationery, postage, fuel, car repairs, servicing and cleaning, insurance, uniform, also rent, rates, lighting and heating used in connection with the business. Where the business is run from one's own home, a proportion of those domestic costs may be set against tax, but advice should be sought from an accountant on this matter in order to avoid complications involving capital gains tax should you wish to sell the property.
- Costs of any books or materials bought for resale; also minor items such as tools, typewriters etc for driving school use.
- Wages and salaries (other than your own). If your spouse does sufficient work to justify a salary, and providing it does not exceed the personal tax allowance when combined with income from other sources, her salary will be tax free.
- Hire, hire purchase and leasing charges. (The rental only is allowable.)
- Interest on business loans and overdrafts.
- Business insurance and professional subscriptions to trade associations.

- Travel expenses associated with the business (not to and from work and hardly applicable to the driving school).
- Bad debts, providing the debtors are named.
- Outstanding trade debts owed by you at the end of a trading year are classified as income and those owed to you are counted as costs.
- VAT on car purchases.

Income tax Schedule E—Pay As You Earn (PAYE)

An employer has a responsibility to deduct income tax from the pay of employees, regardless of whether or not he is instructed to do so by the local taxation officer. If he fails to deduct the tax due then the employer, and not the employee, is liable for its payment. Non-collection may also incur other penalties. The Inland Revenue supplies a free booklet, *Employer's Guide to PAYE*.

To operate PAYE the employer requires:

Deduction cards
Code tables which list the tax allowances attributable to each employee
The tax tables
The blue card P8.

These are all obtainable from the local tax office, to whom deductions are forwarded monthly.

New employees will need to provide you with a form P45 from their previous employer. It gives their code number, salary and tax paid to date.

Each year, a summary of tax deductions and National Insurance contributions for all staff is sent to the Inland Revenue on form P35. Form P60 is completed for each employee at the end of the tax year, showing total pay and tax deducted during the year.

National Insurance

Every employee earning more than £34.00 a week pays a weekly contribution to the social security system, and the employer makes a payment for each of his employees. The NI deduction tables are supplied by the Department of Health and Social Security, and the joint contributions from employee and employer are paid to the Inland Revenue.

Holidays, retirement and ill-health

Being self-employed means no one else is going to pay you wages or salary while you are on holiday. Ensure that you set aside contingency funds for this and remember that bills still keep coming in during your absence.

You will probably not want to work until you can go no further. Provision should be made for your retirement! It is also worth making some provision for accident and sickness. Specialist schemes to cover all these contingencies are available from the Driving Instructors' Association and any other reputable insurance company.

Chapter 5
Marketing Your Services

Before attempting to launch your new business you must consider the factors which influence the selection of a particular driving school by its potential customers and why those learners who choose alternative methods of learning do so.

Financial considerations

Price will be a restrictive factor to potential customers who cannot afford a driving school and do not have access to alternative private facilities.

In many cases, where the potential customer is unaware of any significant difference between the services offered by one school and another, the price is likely to be the determining factor. In other cases, however, a potential customer may pay a higher price for an inferior service because he assumes that a higher price means a better quality service. But price is not the only factor that influences the choice between schools.

Motivational factors and attitude

The attitude of a potential customer is influenced by three fundamental components, namely knowledge (experience and awareness), the emotional characteristics of the individual, and motivational factors which influence behaviour.

The knowledge possessed by the potential customer can be thought of as representing the logical side of the buying decision, such as the price of lessons and value for money in relation to the service offered, and an awareness of the choices available to him.

The decision to learn to drive will be emotionally stressful to some. Motivation to achieve a desired result or to avoid any unwanted outcome induces the potential customer to behave in a certain manner or to take a particular course of action. These decisions and actions express the innermost desires and fears of the individual.

When a customer chooses to undergo a course of driving lessons with a particular school, he is not buying the lessons, but the results which he believes can be obtained by that course of action. The most obvious desired result will be to pass the driving test and obtain the benefits of holding a full driving licence.

It is important to keep this idea of selling the results and benefits of a course of lessons in mind because they are what your potential customers are seeking. The results and benefits to the client may be either emotional or material and include:

Emotional results
    Status
    Confidence
    Personal satisfaction
    Security
    Pride of achievement
    Pleasure
    Peace of mind.

Material benefits
    Passing the test
    Increased safety
    Independent means of travel
    Greater convenience
    Increased income
    Value for money
    Increased leisure time.

It is also appropriate here to emphasise that the decision to purchase a course of lessons is influenced by a desire to avoid unwanted results such as failing the test, fear of ridicule, lack of self-confidence, and the financial consequences of wrecking the family car.

In addition, it is important to recognise that the potential customers' personal attitudes and values will be strongly influenced by social images and family pressures, and that good public relations emphasise the desirability and benefits to be obtained.

Marketing considerations

Marketing involves the identification of consumer needs, the provision of the service, the appropriateness and presentation of the tuition content, and the manner of delivery. Remember, the customer buys the results of the service you provide and these will vary in importance from customer to customer. Have you identified the results of your service? Do they satisfy the needs of your clients? How does your service compare with those provided by your competitors?

Selling the service

A high proportion of your potential customers think that, apart from a few trivial points, all driving schools provide the same basic service. This is far from the truth and if you think the same perhaps you ought not to be setting up your own school.

The service you provide should be unique. In order to exploit this you must identify the various elements which make it so for it is these elements which produce the benefits or results required by the customer. They may be tangible or intangible; examples of tangible aspects are:

The course objectives and syllabus
Your clearly defined and documented training programme
Learning notes for guidance
Appointment cards
Student record cards
Student assessment forms
Modern tuition vehicle fitted with dual controls
Personalised pick-up service
Provision of additional facilities, eg, emergency handling techniques, motorway training,

economy/defensive driving courses, workshop/ classroom facilities, pre-test lectures.

The more intangible aspects of your service will help to provide a back-up to the general service offered and include:

Your reputation and integrity
Your experience and qualifications
Your reliability and consistency
Your general manner and attitude
Free advice and peripheral services
Presentation of the vehicle
Provision of professional reception services
Provision of a 24-hour telephone answering service.

They help to achieve a particular result and can be used in advance to persuade potential clients that the results they want can be fulfilled by you.

A marketing plan

Having acquired the fundamental skills and qualified as a Department of Transport Approved Driving Instructor, you may be, or feel you can be, one of the best driving instructors in Britain. Probably your skills and aspirations in this field have prompted you to consider launching your own school in the first place. Your belief in yourself and your commitment to the project, combined with flexibility and flair, will all affect your ability to survive and prosper.

In addition, however, it is becoming more vital now than ever for new instructors to recognise that the skills involved in attracting potential customers to use their services goes far beyond their expertise in driving instruction. The service you provide must be publicised and brought to the notice of potential clients. They must have a need for the service and it must be presented to them in a manner which will attract and interest them.

Having achieved all of this, your service must then satisfy or exceed the expectations of your customers. Their satisfaction is directly related to your skill, manner and expertise as an instructor. It will, in time, bring your services to the attention of others through recommendation. There are numerous other factors which influence the opinions of clients and these will include your attitude, personality, manner, dress, hygiene, your reliability and even your choice of vehicle.

Setting up a new driving school requires neither a large sales staff nor a complex marketing strategy. However, unless you intend to blunder along as many other driving schools do, you must give serious thought to the marketing aspect of the business and develop a realistic plan on how you are going to persuade potential clients to use the services of your school as opposed to those offered by others in the area. The following is an outline 'blueprint' of such a plan:

Initial Organisation and Planning of Courses

> Make Yourself Known to
> Potential Clients

Advertising

> Arouse an Interest in the
> Services you Provide

Promotions

> Create a Preference for Your Services
> over those of your Competitors

Recommendations

> 'Book Up' the Client
> Retain the Business and
> Obtain Future Recommendations

Market research is one of those terms which frighten most instructors who don't see the need for it, and consider it's mostly common sense that business comes from recommendations by satisfied clients, so why should they bother?

This attitude probably explains, in part anyway, why most driving schools are one- or two-car operations working from home. Most proprietors drift along totally oblivious of the tide, weather and currents influencing their progress, or otherwise, in the rough sea of driving school competition.

Some basic information about potential clients is crucial to the consistent success, prosperity and growth of any business. This involves a knowledge of the market size and how much is spent on driver training and traffic education every year.

What are the overriding trends influencing the market? For instance, the peak booking periods for new clients is February to May and August to October each year. July and August are usually the quietest months for trade because most people go on holiday during this period. Christmas is a bad time when cancellation of lessons reaches its peak as customers find themselves short of time and cash, particularly in the week between Christmas and the new year, making this a good time to holiday yourself.

November, December and January may be bad months to start up the business and so are May and June. These are usually the slackest months for new clients coming in which makes it more difficult to build up quickly.

Of course, it largely depends on the type of school you are envisaging and what particular part of the market you are aiming at, but these will require some consideration at the outset.

A knowledge of your competitors is also essential. Who are they? What type of services do they offer? How do they arouse interest and create a preference for their services? Visit their premises and have a look! See what they are offering their potential clients. Where do they advertise their services?

Suppose you wanted to book lessons for your son or daughter—where would you look up the local driving schools? Who would you be inclined to book with? When you have answered these two questions, try to analyse the answers and use the information for yourself.

Where is the biggest market for driving lessons?
By far the biggest and most established market is the
unlicensed driver who wants to pass the driving test
and there are another 19,999 instructors competing
for it. There are 2 million provisional licence holders
at any one time and all you must decide is how you are
going to reach those in your area in order to get your
message across.

We know certain things about them; for example,
we know that the vast majority of people learning to
drive are under 25 and many of these are students.
This information is important because already it is
starting to narrow the field down and this helps you
to make the most cost-effective use of your publicity
and advertising.

We also know that a fairly high proportion of people
wanting to drive make enquiries and book their lessons by telephone; most others book personally at the
driving school office. (It should be emphasised that,
although most driving instructors operate from
home, these are almost without exception one- or two-
car schools; to consistently sustain sufficient business to run three or more cars, a central office will be
required.)

Already you can see we have isolated a large proportion of the market and identified some of their
habits. For example, you can see that a properly
worded letter to the principals of local schools and
colleges outlining your services could find its way to
the student notice board. A similar letter addressed
to the students' union could have the same effect.
Youth leaders too may promote your services in this
manner. Where do young people spend their leisure
time? The local cinema, disco, sports club etc.

Let's not forget the rest of the market, however.
There are well over 28 million licensed drivers in
Britain and many of these would benefit from further
training. These benefits would be financial, increased
safety, improved status, confidence and satisfaction.
Industry needs to be convinced of the savings to be
gained from economy and defensive driving courses
provided for staff, the costs of which can be recouped
over and over again through savings from lower

repair bills, reduced insurance premiums and more efficient drivers. A letter to the personnel officers of large companies outlining the benefits of your services to both employers and staff could find its way to the works notice board.

Most people book driving lessons by telephone, making an insert in the Yellow Pages essential. It may seem expensive, because you pay a year in advance; however, you usually get more than a year's exposure before the directory is replaced! This is a far more effective way of advertising yourself than in the local paper at, say, £10 per week and at around half the cost. The only problem with Yellow Pages for a new business is that it probably takes a while before the new directory with your name in is distributed.

Chapter 6
Advertising and Publicity

Large companies like to project a clear image of their product or service and put all their efforts into developing and promoting that image. When starting up your new business try to emulate this example. The purpose is to develop an identity, consistency and the appearance of an established firm which knows its business.

Corporate image

Planning and developing this image starts with the name, which should describe the nature of the services you are offering. Establish the business style and logo and stick to the same design and format for all stationery and advertising.

It is advisable to seek professional advice and guidance in the design of a logo and style at the outset so you don't find you have to change midstream. This may cost more when you can least afford it but is well worthwhile if you get it right first time. A professionally produced collage can illustrate visually the services you provide and these are more interesting to look at than type.

Professionally designed artwork will be the best— newspaper art departments are not particularly good at designing attractive and interesting advertisements and millions of pounds are thrown away every week because of this. You will need to get a high response for every penny spent on your initial advertising campaign and the image projected so the material must have a good impact on the reader.

Advertising

If you are to succeed and prosper as a driving school you must, particularly in the early days and in the absence of a central office, advertise constantly. It is not sufficient to advertise once or twice and then sit back waiting for the business to roll in.

The public, or more specifically your potential clients, are fickle. They are also very forgetful and have a large number of driving schools to choose from. They need to be frequently reminded who you are and what you have to offer.

Most advertising will be, by nature, informative. However, it should not be assumed that because of this it will be persuasive and generate the desired response. You should aim to build up an active response from the recipients of your advertising campaign. For example, you will, at one end of the spectrum, want to persuade potential clients simply to book their lessons with you, and at the other to have a more positive attitude to the services you provide.

For effective, professional results, it may sometimes be appropriate to approach an advertising agency with a view to assisting you with the design and placing of advertisements. Most agencies derive the bulk of their income from commission paid by the media for placing advertisements in their publications. These agencies can provide expertise in creative design, presentation and the wording of ads.

Obviously you will have to pay more for this type of service and in some instances, where new businesses are spending only relatively small amounts on the actual advertising, it may make the cost of their services seem unrealistic. It is, however, an area which should be explored and you can only learn from a meeting with such a company, even if you choose not to avail yourself of its services.

Do not commit yourself to an advertising programme which you are unable to support financially or which is unrealistic in terms of your needs.

One of your main priorities upon launching your new school will be to bring it and the services provided to the attention of potential customers. You will want to do this quickly and as cost-effectively as possible, without putting too much strain on your capital resources. You will also want to present the new business in a favourable light to other public groups which will include parents, teachers, employers and others linked in some influential way to either your potential or existing clients or to your business.

The main purposes of advertising is either to boost sales by persuading potential clients to book their lessons with you, or to encourage a favourable attitude towards your business, which is designed to increase confidence in it, establish an identity or to promote a change. Advertising is a tool to be used in the correct place at an appropriate time and for a specific job. If you are starting up a local driving school there is little point advertising in the national press.

One of the first priorities is to select the correct medium through which to advertise your services. If you advertise in a local daily paper it will pay you to find out which days provide the best response by monitoring the results.

It may also be worthwhile to experiment with the wording, but stick to the same theme—use the one which provides the best results and which will become recognised as your own house style.

Don't put too much information in; if you do people will tend to ignore it. Be precise about the services offered. Make sure your name and telephone number are prominent.

Keep a check on where most of your customers are coming from. This will enable you to concentrate future efforts on the media providing the best result.

Choice of advertising media

There is a wide choice of media by which to advertise your services. Some of these will be totally inappropriate to the new driving school starting up in a small local area. For example, the national press, television and national motoring magazines are appropriate only to national organisations and products with a national distribution potential. Although these can be the most effective ways of advertising a new product, it is obviously wasteful of resources to generate business in areas you do not wish to cover. Be realistic about the areas you do cover and concentrate on reaching those potential clients nearby, at least initially.

Whichever method of advertising you choose, each will have its advantages and limitations.

Local press. No other publication or news medium comes near local newspapers for credibility, interest and the thoroughness with which they are read by the local community. They range from the local daily, with hundreds of thousands of readers, to the small weekly covering only the neighbourhood.

Local people identify themselves very strongly with their own local paper because they love reading about people they know and there is rarely any item of business news that won't attract the editor's attention, providing there is something newsworthy about it. Locals include:

Weeklies
Evening dailies
Free weekly trade magazines
The parish magazine
Club newsletters
Supplements
Free sheets
Immigrant newspapers
Local magazines
Student magazines
Trade union magazines or inserts.

Dailies. Although local, these will normally cover a vastly greater area than you can realistically serve when first starting up, which can make a daily paper an inefficient and wasteful form of advertising. On the other hand, if you are operating from a central office in a large town, it can be an effective way of letting a new clientele know that you exist and where you can be found.

Weeklies. The coverage of the very small local weekly is about the same area as you can comfortably service when operating a personalised pick-up service. If you are located on the fringes of the readership area, you may have to consider the benefits and disadvantages of advertising in two or maybe more papers.

This sort of publication covers a fairly high although localised readership, it is relatively cheap and can be booked easily and quickly. The readership is, however, non-specific and there will be many with neither the need for nor the interest in your services.

Local community magazines. It is relatively easy to identify the readership of such magazines and newsletters. This will make it easier for you to direct your advertising to those most likely to respond; there are, for example, student magazines and similar, local hospital newsletters and a vast array of other materials put out by local clubs and youth centres.

Advertising in this type of publication is relatively cheap. If you have the facilities, it may even be a worthwhile proposition to produce short newsletters etc for local organisations at cost in exchange for free advertising in them. Such small organisations are always short of cash to finance their administration and will often welcome such help. It also provides an opportunity to be seen to be doing something worthwhile for the community and to get talked about. In some cases, where you are unable to provide a free service because of the work involved, it might be possible to provide it at cost or near cost, depending on how much work is involved.

Leaflets/Inserts. These can be an effective way of introducing yourself to the local community. The material has only a short life but the response is quick. They are also easy to direct into the areas in which you operate. Distribution might be carried out by yourself or family or through local newspapers combining them with the delivery of their publications. Alternatively, local firms exist in most areas who can arrange deliveries for you.

Door-to-door distribution. The costs for this service range from about £15 to £20 per 1000 copies delivered. Coupled with some kind of promotion, this could make an effective way of introducing your services when combined with the initial launch of the school.

Posters. Attractively designed posters placed in local filling stations, post offices and other shops are another useful means of reminding the local community who you are and what services you provide. Small commercial poster sites are available inside and outside buses, at bus stops and in outer shopping areas (although problems may arise if the bus services cover too large an area or if local companies allocate buses to new routes).

Posters can have a high impact, a fairly long life and are relatively cheap. The message should be short, sharp and easily read. Don't expect people to queue up to read the small print though!

Telephone and business directories. Yellow Pages and Thomson directories are relatively cheap, have a long life and provide a constant source of new business. The major problem for the new school, however, is that infrequent (annual or longer) distribution of these can cause considerable delay in getting known and wherever possible this type of advertising should be planned well in advance.

A very high percentage of driving school clients refer to Yellow Pages and similar directories. The message to be delivered should be short and to the point.

Local radio. Commercial radio stations can be another outlet for reaching potential clients. The production costs of a sufficiently thorough advertising campaign may be prohibitive, however, for the new school.

Local cinema. This can provide relatively cheap exposure time but production costs can be high. The audience is primarily the young and as such is worth consideration.

Display advertisements
These can be used to display your services in an attractive eye-catching style designed to generate interest in your service and to persuade potential

customers to book lessons with you. They should provide the customer with all the information needed, such as name, address, telephone number and details of the services provided.

Within the limitations of the publication dimensions and your own pocket, the advertisement can be any size you choose. Rates are usually calculated per single column centimetre, often with special rates for full, half, quarter or one-eighth pages.

Classified/Semi-display advertisements

These usually appear under a classified heading, often near the back of the publication if it is a local newspaper. Some publications, such as Yellow Pages or Thomson directories, are classified throughout. Classified headings are usually Driving Schools, Motoring Services, Driving Tuition, or Schools of Motoring.

Classified ads need not be small ones and display ads can often be placed in classified sections.

Semi-display ads are usually too small to appear in other sections of a publication and so are grouped together in the classified section.

Feature advertisements

These are particularly used by publications which derive all their income from advertising, such as free local weekly newspapers. You may be able to persuade publishers to give you a free editorial story in exchange for an advertising contract.

Other publishers also provide these 'free sections' from time to time and include them as inserts in an existing newspaper or other publication. The general idea is that they have a single theme such as motoring and encourage advertisers by offering special rates and possible editorial mention within that section.

The sales letter

This is perhaps one of the most important means of advertising your services at low cost and it can be very effective. One letter to all the major employers in an area could find its way on to company notice boards.

The essential ground rules when composing such letters are:

1. The letter should be single-page size, or even less, on A4 headed paper of good quality.
2. It must look professional.
3. It should be addressed personally (by name) to the managing director or personnel officer. (A brief telephone call to the company concerned should give you the required information.)
4. The letter should start with a strong headline in bold or underlined print. It should take no longer than four or five seconds to catch the reader's attention; the letter, therefore, should be kept short or it may end up in the waste paper basket.
5. State immediately how your services can benefit the reader, the company and its employees.
6. Keep the letter flowing with a friendly, helpful tone as if you were actually speaking to the reader. Use the word 'you' wherever possible, remembering there is another human being on the receiving end. Reassure the reader that you understand his problems. Don't refer to what he can do but emphasise the benefits of your services.
7. Encourage immediate action, tell the recipient what to do with the letter or enclosed material, eg, post it on the staff notice board.
8. Where a reply is required enclose a pre-paid reply card or envelope.

   Be sure to repeat the process with local schools, hospitals, student unions, colleges of further education etc, remembering in each case to relate the letter directly to the organisation concerned.

Special promotions
Special promotions are normally designed to create a rapid but short-lived increase in business. As part of a long-term, planned campaign they can permanently increase business. They are often used to launch new

products or services and driving instruction need not be an exception.

When considering what type of promotion is likely to be the most effective you must decide clearly what you are trying to achieve because there are various methods and techniques available. The following may be appropriate to the driving school.

Free demonstration lesson. The principle is that once the client has committed himself for the free demonstration he will, almost without exception, carry on having lessons at the normal rates.

Special rates for courses of lessons. A reduction is offered for courses of lessons paid in advance.

Special lessons. Extra time can often be given on a first lesson.

Coupons/Vouchers. These can either be delivered through the door or cut out of adverts offering limited cash discount or a 'free' lesson in exchange for the voucher.

Special demonstrations. These may take the form of special film evenings for pre-driving test candidates or group pre-driving sessions to reassure and motivate potential customers. They provide the instructor who has the appropriate skills with an opportunity to display and exploit them. These sessions should not be attempted without the prerequisite skills in group training/public speaking.

Special discounts. They are often offered to company employees, students and others.

Special time discounts. Offered for lessons taken during slack times of the day to increase overall efficiency of the business.

Guaranteed results. Money-back guarantees should not be offered. Even with the best of students it is too risky.

The results of promotions should be carefully monitored to check that the desired results are being achieved. Large amounts of money can be lost by making silly offers which have either been miscalculated or go wrong.

Public relations

Public relations is a greatly neglected business tool. Untold time and effort are devoted to improving skills in finance, sales, personnel management, stock control, advertising and the many other subjects which together make for a successful business, yet for some reason, the very term public relations leaves many businessmen cold. Moreover, some of those who are aware of it are guilty of approaching it the wrong way, using public relations to cover up faults rather than for establishing positive relations with the public.

Be Your Own PR Man, written by Michael Bland, takes the myth and mystique out of public relations. It shows the small businessman how to attract customers, increase sales and achieve better publicity by using the techniques of PR, often at little or no cost. It argues that PR is not an expensive corporate operation, but a way of achieving effective and favourable publicity for business ventures of all sizes.

The author covers the basic principles of public relations and explains how to deal with the press, television and radio. The book gives practical information and detailed advice on the methods involved in writing a press release, getting coverage in the media, conducting and publicising a survey, giving a speech, engaging in community relations schemes, sponsorship events, and so on. Many ideas are discussed in this book but, once you have established the ground rules, PR is whatever you want to make it. It's the difference between total anonymity and people knowing that you're there; between 'plodding along' and 'taking off'.

Press releases
Local newspapers like to write about local people! Anything which is new or different from the normal theme will be of interest to someone. Let the press know that you are prepared to comment upon local

and national issues concerning motoring, one-way systems, accident blackspots, new legislation etc.

What is new, different or interesting about the services you propose to offer? For example, you may intend running your business differently from the traditional practices in your area, or offering a completely new type of service to the local community.

Motoring is a very topical subject with wide appeal and new laws affecting motorists are introduced fairly regularly. You may wish to comment on these or other motoring events. Contact your local reporter and offer your thoughts, or better still try to develop a working relationship with him on aspects of local motoring interest. Don't expect or wait for the press to come to you and don't forget your local radio stations.

When being interviewed by the press it is important to get your thoughts clearly together beforehand. Be sure your facts, figures and other information are correct. Decide beforehand which main points you want to get across. Restrict these to three at the most and jot them down. The same rule applies to press releases: keep them brief and to the point.

It is as well to consider at this point what makes news. It's what people are talking about and doing, and events and decisions which affect them. In deciding what is newsworthy journalists look for hardship or danger to the community, unusualness and novelty, scandal and conflict, individualism and outspokenness, in any combination.

Not all news, however, falls within these categories and a basic service provided by local weeklies to the community is about births, marriages and deaths and the human stories of success and failure which occur between. You need only glance through the papers to see this for yourself. Most of the stories you read do not get there because the reporter digs them out for himself. He is told by people like yourself who have something to say. It may be the appointment of a new managing director by a local firm, an Oxford scholarship awarded to a local student, the opening of a new sports ground in the town, or even the new driving school with a novel view on how things should be done.

The press release might be used for a number of reasons, for example to:

Give advance warning of a celebrity opening a new driving school

Provide a report or opinion on the trends, events and decisions which affect the motorist or new driver

Announce special promotions and events

Provide a report on items of significance, progress and success

Provide general background information on new projects

Announce the introduction of new services or offices

Announce the appointment of new staff.

A press release should follow the principles of *who, what, where, when* and *why*.

The Driver School of Motoring	WHO
is opening a new office	WHAT
at 31 High Street, Centre Town	WHERE
on Saturday 1 March 1984	WHEN
to satisfy increased demand.	WHY

I. Ambest of Driver Motoring School	WHO
was appointed manager	WHAT
of the Centre Town branch	WHERE
on Saturday 1 June 1984	WHEN
after his recent award of the Diploma	WHY
in Driving Instruction by the	
Driving Instructor's Association.	

In each case the WHY can be extended as a natural follow-up article.

Chapter 7

Expanding the Services Offered

In spite of the strong competition within the driving school industry, opportunities still exist for the extension and expansion of existing services and the provision of new ones. There is scope for improving the range of services currently available to the 'L' driver. These include improving the efficiency of existing training through the introduction of properly structured courses and group training facilities both for in-car and workshop activities. Most areas of the driver instruction/traffic education market are listed below. Some of them are still largely unexploited.

Courses for unlicensed drivers

Traditional driving lessons: one-to-one in the car
Full-time intensive courses
Group training in the car
Group training in the classroom
Pre-driver courses
Contract services to industry and others
Adult education contracts
Disabled driver training
Conversion courses for motorcyclists
Conversion courses from three-wheeled vehicles
Motorcycle training programmes

Courses for licensed drivers

Refresher courses
Emergency handling techniques
Advanced driving lessons
Defensive driving courses for industry
Motorway driving techniques

Caravan towing techniques
Motorcycle training programmes

The areas offering most future potential are:

Introduction of properly structured training
programmes
Pre-driver courses and traffic education
Full-time and residential courses
Defensive driving courses for industry.

Structured training programmes

Newly qualified drivers continue to be over-proportionately involved in road accidents. This focuses permanently on the need to improve the quality of driver education programmes and scrutinise existing training practices.

The main aim of any driver training programme should be to minimise the risk of accident involvement. Modern teaching strategies should be used to produce safer behaviour patterns rather than pursuing some traditional practices of automatic repetition of physical skills and simple recitation of rules and regulations. Learning only how to cope with a dangerous situation is insufficient to produce desirable behaviour.

Training should be properly structured and include practical experience in how to recognise, avoid and prevent the hazardous circumstances which lead to conflict. There is clear evidence to show that most conflict is caused by deficiencies in the visual scanning and hazard recognition processes rather than lack of vehicle control.

It is well established that perceptual abilities are limited and drivers frequently, and mostly unknowingly, expose themselves to an information overload. The effect of this can be compared with a temporary fuse in an electrical circuit. Faced with this perceptual overload, the driver may only attend to some of the problems he is facing while completely rejecting and neglecting the rest. The driver must learn to recognise this phenomenon and be prepared to change the circumstances which lead up to it.

The long-term benefits of the implementation of such a programme are immense when it is considered the cost of road accidents to Britain is currently approaching £2000 million per annum. Low-risk driving strategies have already proved they can reduce these costs by up to 50 per cent. The cost to the learner driver need be no more than for using traditional training practices, and could be less.

'L' driver practical training
The traditional way of taking driving lessons is usually one or two sessions per week over a prolonged period of time. This helps to spread the cost but greater efficiency could be achieved by the use of more condensed, full-time courses.

Group learning is another way of reducing the cost of learning to drive while at the same time improving standards. Drivers can be sensibly divided into a beginner's course, followed by individual instruction, and finally a pre-test course. Students of similar standard could be trained together using such a combination of individual and group training to a higher standard at lower cost.

Industry and the armed forces require training for the 'L' driver and there should be opportunities for the keen instructor to negotiate contracts with these organisations where they are fairly local. The Manpower Services Commission, a government sponsored body, frequently requires driving instructors for training students on some of their courses where driving will be a necessary part of the subsequent work.

Full-time and residential courses and more concentrated tuition

There are sound educational reasons why learning to drive should not follow the traditionally extended process. One lesson a week for anything from six months to a year, and in many cases much longer, is a very inefficient way of learning, estimated to increase the number of lessons required in some cases by more than 25 per cent. Full-time intensive courses are becoming more popular, and worthy of greater consideration by instructors.

Difficulty may be experienced in getting driving test appointments and there is also the possibility of the client not being up to the required standard at the end of the course.

The advance booking of tests can largely overcome such difficulties and alternative arrangements can normally be made where the client may be unsafe to take the test.

A financial service applicable to such courses is available from Avco Trust, Avco House, Castle Street, Reading RG1 7DW; 0734 586123.

Customer credit

If you wish to arrange loans for potential customers, perhaps for the purpose of offering condensed, full-time or residential courses, you will require a credit licence under the Consumer Credit Act 1974. These are obtained by applying to the Office of Fair Trading, Bromyard Avenue, London W3 7BB. To obtain a licence you must be a fit and proper person to be in the business. If you are a reputable person you should have no difficulty in obtaining one.

'L' driver in-class training

Some aspects of driving can be learnt just as effectively, and probably more so in some cases, in a classroom rather than in the bustle of traffic. By and large the public are a little sceptical of classroom education where driving is concerned but there are still many who are eager to attend classes provided either by a driving school or adult education services (the latter being a sound source for new pupils for the practical side of their tuition).

In-car and in-class courses are held by some education authorities and there are many varied opportunities for instructors in this field both in state and private schools. Driving instructors are advised to attain suitable training and qualifications before embarking on any schemes involving classroom teaching.

Advanced/Defensive driving

There is a fundamental difference between advanced and defensive driving.

Advanced driving is a system of car control and driving techniques based on the experience and methods developed by police drivers. These methods and techniques are based on the concept that a vehicle should always be in the correct position in the road, travelling at a proper speed for the prevailing conditions and with the correct gear engaged.

The overall philosophy works well and 'advanced' drivers are known to have a better safety record than others. However, methods of assessing advanced driving are fairly subjective. A recent report showed that Department of Transport driving examiners were more severe in their assessments of 'L' test candidates in some areas than standards applied to advanced driving.

Defensive driving has a more objective approach which embraces the same overall principles as advanced driving but with a significant difference in emphasis in that its sole purpose is accident avoidance.

Advanced driving is based upon skill and experience whereas defensive driving is based upon the identification of risk factors and professional training to minimise these.

You may still feel it is two different ways of saying the same thing. Well, maybe it is and certainly the difference is difficult to define. Advanced driving can be described as being based on *knowledge and skill* whilst defensive driving is based on *attitude*.

Teaching advanced, low-risk and economy driving techniques to experienced motorists can make a refreshing change from learners. Industry needs to be convinced of the savings to be gained from defensive driving courses for its staff and also that the cost of such courses can be recouped through lower insurance premiums, reduced repair bills and more efficient drivers.

Automatic transmission tuition

Services offering tuition on automatic transmission vehicles will find a tremendous in-built public resistance. This may be due to the fact that persons passing the driving test on such vehicles are only entitled to drive automatics. The majority of people do not want this restriction imposed upon them. Automatics, however, are a tremendous help to the disabled driver, people with other handicaps to learning, and those of advancing age who find learning difficult. They help to reduce the demands imposed on the new driver.

The provision of this type of service is not recommended for the 'start-up' school and, in any event, will require an extremely high population density to support it.

Profitable sidelines

As a salesman, the driving instructor is in an enviable position. The provisional licence holder and the newly qualified driver spend millions of pounds each year on:

1. The *Highway Code* and other home study materials
2. Driving books
3. Driving shoes and gloves
4. Sunglasses
5. Second-hand vehicles
6. New vehicles
7. Vehicle insurance
8. Membership subscriptions to motoring organisations.

There is no reason why the instructor should not sell items 1, 2, 3 and 4 in the list, receive introductory commissions on 5 and 6, and act as agents for 7 and 8. After a short time, and for very little outlay, these sales could net £20 extra on the weekly income and after a few years, considerably more for very little additional effort.

95

Diversification and use of existing facilities

Opportunities exist for driving instructors working in cooperation with established businesses in associated fields such as car dealerships, insurance brokerage, motorist accessory shops. In the not too distant future possibilities also exist in the home computer market as software programmes become more readily available.

All parties involved in such schemes should be able to benefit considerably providing the project is organised and planned properly from the outset.

Motor insurance agents with expensive High Street offices might like to share the costs with an associated business such as the driving school. In addition to reducing the costs of running and staffing the office, both businesses cannot fail to benefit providing the scheme is organised properly.

Motor car dealers are always trying to entice the public into their showrooms. In the long term, what better means is there to attract tomorrow's drivers than in exchange for the use of their facilities for group training, film nights, pre-driver sessions and also pre-test demonstrations?

Driving instructors, in association with proprietors of small country or seaside hotels (and vice versa) may be able to organise holiday packages which incorporate a driving course with test at the end.

Partnerships and cooperatives

These may provide a way of reducing the costs of running a central office or of enabling a group of local instructors to open an office and share the costs.

Such partnerships will enable instructors to provide a wider range of services to the public such as some in-class content in the training programme, a more professional booking service for customers; expansion into other fields such as motorcycle training and courses for industry is also much easier as a group than individually.

(Warning! It is inadvisable to go into partnership with anyone you do not know or trust well. All partners are liable for the business debts incurred by others.)

Heavy goods and public service vehicle driving schools

HGV and PSV are quite separate areas of instruction from training learner drivers, and holders of these licences may be considering a career in this field. In view of the complexity of these vehicles and the higher standards of driving concerned, it may sound strange that there is no legal qualification required other than a full licence to drive the class of vehicle on which training is given.

Anyone contemplating starting up their own HGV school, however, should recognise that they are unlikely to be successful without an adequate background and experience of training in this field. Those wanting to start a career in teaching on these vehicles should first successfully complete an appropriate course with the Road Transport Industry Training Board. They should also take into account that the costs involved in starting up such a training school are far greater than for a motor car driving school and to be successful will need to provide classroom training facilities in addition to vehicles. The cost of maintaining HGV tuition vehicles is also extremely high.

In addition, the boom years in HGV training have passed, partly because of the current economic recession and partly for other reasons. It seems unlikely that an upward trend will recur.

Chapter 8
The Register of Approved Driving Instructors

The Department of Transport Register of Approved Driving Instructors was introduced in 1964/65. It was initiated on a voluntary basis to give established, practising instructors an opportunity to take the qualifying examinations in anticipation of proposed legislation requiring registration. The purpose of the Register was to ensure that all persons teaching the public to drive for profit or reward were able to meet minimum laid down standards. These standards and other requirements are specified in the regulations relating to the Register.

In 1970 it became compulsory for any person giving professional instruction in driving a motor car to have their name entered in the Register. Upon the introduction of the compulsory Register it was also necessary to make some provision for newcomers to the industry to obtain experience and training prior to taking the qualifying examinations. The *trainee licence* was introduced to fill this role.

The Register of Approved Driving Instructors and the licensing scheme for trainee instructors are administered by the Department of Transport under the provisions of the Road Traffic Act 1972. The following information summarises the main points relating to the administration of the Register.

Professional driving instruction

It is an offence for anyone to give professional instruction (that is instruction paid for by or in respect of the pupil) in driving a motor car unless:

1. His/her name is in the Register of Approved Driving Instructors; or

2. He/she holds a trainee's 'licence to give instruction' issued by the Registrar.

Application forms for registration or for a temporary training licence can be obtained from any Department of Transport Traffic Area Office (list on page 114). Completed forms, with the appropriate fee, should be returned to the local Traffic Area Office.

Anyone who wants to be an Approved Driving Instructor (ADI) must:

1. Have held a 'full' (ie, not provisional) driving licence for periods amounting in the aggregate to at least four out of the six years preceding the date of the application (any period after passing the driving test during which a provisional licence is held may be counted towards the four years);
2. Not have been disqualified from driving for any part of the four years preceding the date of his application;
3. Be a fit and proper person to have his name entered in the Register. (Non-motoring convictions, motoring convictions including those not resulting in disqualification, providing they are not 'spent' under the Rehabilitation of Offenders Act 1974, are taken into account in assessing an applicant's suitability.)
4. Pass the Register qualifying examination.

Certificate, period and renewal of registration
The name of an applicant is entered in the Register when he has qualified and paid the stipulated registration fee. He is then entitled to receive an official Certificate of Registration which incorporates his name, photograph and the official title. The certificate is suitable for display in the car which he uses for tuition. There is no legal obligation to issue a duplicate certificate but if a registered instructor satisfies the Registrar that the original has been lost or destroyed, he can get a duplicate on payment of the appropriate fee.

The official title of a registered instructor is 'Department of Transport Approved Driving Instructor', and he is authorised to use this title as long as his name remains in the Register. It is an offence for a person to use this description if he is not entitled to it.

Special note: Provisions exist within the Road Traffic (Driving Instruction) Act 1984 which will require professional instructors to exhibit a certificate affixed to the vehicle in which tuition is given and which will indicate that the instructor's name is on the official Register.

Registration normally lasts for a period of four years, and before it expires an Approved Driving Instructor can apply for it to be continued for another four years. A renewal notice is sent to him at his last known address about a month before his period of registration runs out, but if he does not receive this, he should apply to DTT3, 2 Marsham Street, London SW1P 3EB. When registration is renewed, it is not necessary to take the qualifying examination again unless there has been a lapse of a year or more since the expiry of the last period of registration, but the instructor must have the other qualifications for registration and must not have refused to undergo a test of continued ability and fitness to give instruction or failed to reach the accepted standard in that test.

Test of continued ability and fitness to give instruction

When required to do so by the Registrar, an Approved Driving Instructor must undergo a test of continued ability and fitness to give instruction. The test includes the attendance of the examiner while the instructor is giving a driving lesson to a pupil. The instructor is assessed on the method, clarity, adequacy and correctness of his instruction; the observations and proper correction of his pupil's errors; his manner, patience and tact, and his ability to inspire confidence. The tests are conducted during the examiner's normal working hours.

An Approved Driving Instructor may have his name removed from the Register at any time if the Registrar is not satisfied that he still possesses the qualifications for registration, or if he refuses to take a test of continued ability and fitness to give instruction or fails to reach an acceptable standard in that test.

Licences to give instruction

If a person wishes to acquire practical experience in giving instruction with a view to undergoing the qualifying examination for admission to the Register, he may be granted a licence. Licences are issued only for this limited purpose. They are not an alternative to registration and it is not essential to have held a licence before becoming registered. Licences are granted only to applicants with the necessary qualifications (listed below) and are issued subject to conditions. Licences are valid for six months, and a trainee instructor is not normally granted more than two.

The qualifications required by an applicant for a licence to give instruction are:

1. He must have held a 'full' (ie, not provisional) driving licence for periods amounting in the aggregate to at least four out of the six years preceding the date of his application (any period after passing the driving test during which a provisional licence is held may be counted towards the four years).
2. He must not have been disqualified from driving for any part of the four years preceding the date of his application.
3. He must be a fit and proper person to have his name entered in the Register. (Non-motoring convictions, motoring convictions including those not resulting in disqualification, providing they are not 'spent' under the Rehabilitation of Offenders Act 1974, are taken into account in assessing an applicant's suitability.)
4. He must apply to take the written part of the

101

Register examination when applying for a licence unless:

(a) He has already taken the written part and is waiting to hear the result; or

(b) He has already passed that part of the examination within the preceding three years.

Special note: Provisions exist within the Road Traffic (Driving Instruction) Act 1984 which will, in the near future, require new instructors to pass the written examination and the test of driving technique before a trainee licence can be issued. It is anticipated that these will come into force during 1984.

The new regulations may also require some formal training in instructional techniques to be a precondition of a trainee licence being granted. (This provision is still under discussion by the Department of Transport and driving instructors' associations.)

Licence conditions

The conditions under which a licence is granted are that:

1. The holder is authorised to give professional instruction only from the address specified in the licence.

2. There must be at least one Approved Driving Instructor working at the specified address, for every licence holder employed there.

3. For the first three months of the period for which a licence is in force, the licence holder must be under the direct personal supervision of an ADI for at least one-fifth of the time for which the holder gives driving instruction under the licence.

   'Direct personal supervision' implies a close measure of control of the trainee's work by the supervising ADI. For example, a preliminary briefing before the trainee gives a lesson to a pupil would not count as direct personal supervision; the supervising ADI must accompany the trainee during the lesson. This condition does not apply to the second of two consecutive licences.

4. The licence holder must keep a record of the instruction he gives and the supervision he receives during the three months; details of what is required are given below. The completed record must be signed by the licence holder and countersigned by the supervising ADI; it must be produced on demand to an authorised officer of the Department of Transport, and kept for six months unless previously surrendered.

The Registrar may revoke a licence if:

Any of the conditions subject to which it was granted are not complied with; or
At any time since it was granted where any of the conditions referred to previously are not met; or
The licence was issued by mistake or procured by fraud.

Details of trainee's records
The record which a licence holder must keep of the instruction he gives and the supervision he receives should be clearly set out, legible and unambiguous. It may be kept either in typewritten or manuscript form. No specially printed forms are available from the Department for this purpose. The record must contain the following particulars:

1. The name of the holder of the licence
2. The licence number
3. The name of the establishment from which the holder of the licence has given instruction
4. The name of the person under whose direct supervision the holder of the licence has given instruction
5. In respect of each working day:
   (a) Date
   (b) Total number of hours spent giving instruction
   (c) Periods spent under supervision
   (d) Signature of the holder of the licence
   (e) Countersignature of the supervising instructor

Trainee Licence Holder's Record
Driving School/Training Establishment . . . .
Trainee Licence holder's name . . . .
Licence no. . . . . Date issued . . . .
Named Supervisor/Staff Instructor . . . .

Date	Daily hours of tuition	Total hours tuition	Daily supervision	Total supervision	Trainee signature	Supervisor signature

Appeals

Before refusing to renew, or deciding to terminate, a driving instructor's registration, the Registrar is required to notify the instructor of his intention and the instructor is entitled to make representations to him within 28 days with a view to reversing the decision. If the Registrar confirms his decision, the aggrieved instructor has the right to appeal to the Secretary of State within a further 28 days.

Before refusing to renew, or deciding to revoke, a licence to give instruction, the Registrar is required to notify the licensee of his intention and the licensee is entitled to make representations to him within 14 days. If the Registrar confirms his decision, the licensee can appeal to the Secretary of State within a further 14 days.

On receiving an appeal, the Secretary of State appoints a board to hold an inquiry into the whole matter. The appellant can present his own case at the inquiry or be represented by a solicitor, counsel or any other person. The Secretary of State decides on the appeal after considering the board's recommendation. He also has power, if he so decides, to charge the appellant with the cost of the appeal.

An instructor can appeal to a magistrate's court (sheriff in Scotland) if he thinks that either part of the examination was not conducted in accordance with the regulations.

Further information about the procedure on making a formal appeal is given in a leaflet which can be obtained on application to the Registrar.

Registration, licence and examination fees

A fee is payable for admission to the written part of the Register examination and successful candidates are required to pay a further fee before taking the practical part. Subsequent attempts at either part of the examination must also be paid for. After a candidate has passed both parts and before his name can be entered in the Register he must pay a registration fee, and every four years thereafter a fee is payable on applying for renewal of registration. Applicants for a licence to give instruction must also pay the appropriate fee. Since 1 October 1982 the fees have been as follows; details of any changes can be obtained from the Traffic Area Offices.

Written examination	£25
Licence to give instruction	£45
Practical test	£60
Fee for registration	£50
Four-yearly extension of registration	£60

Where a candidate fails to keep his appointment for either the written or practical part of the examination he will forfeit the fee paid unless he has given the office which made the appointment at least three clear days' notice (day of receipt, day of examination, weekends and public holidays excluded), of his inability to attend. Only in very exceptional circumstances will the Department be prepared to consider the refund of a fee where less than three clear days' notice is given, or where the examination does not take place or is not completed for reasons attributable to the candidate or any vehicle he provides. Fees for registration or for licences issued are not refundable.

Department of Transport
ROAD TRAFFIC ACT 1972
Application for Registration in the Register of Approved Driving Instructors

ADI 3 (revised Nov 1981)
Reference No

1. Surname .... Mr ☐ Mrs ☐ Miss ☐ Ms ☐ *

Full christian or other name(s) ....

2. Private address ....

3. Postcode ....

4. Telephone No ....

5. Have you previously applied for registration as an Approved Driving Instructor? Yes ☐ No ☐ *

6. Do you hold a current substantive (ie not provisional) licence, issued in Great Britain or Northern Ireland, to drive a motor vehicle? Yes ☐ No ☐ *

7. What is the Driver Number on your driving licence? ☐☐☐

8. Have you held a full driving licence, or a foreign licence**, for periods amounting in the aggregate to at least 4 years in the last 6 years? (after passing driving test any unexpired portion of a provisional licence counts towards required 4 year period). Yes ☐ No ☐ *

9. (a) Except for spent convictions †, have you ever been convicted of a motoring offence? Yes ☐ No ☐ *

   (b) If YES, give particulars of each offence as follows (continuing on a separate sheet if necessary).

Particulars of offence	Name of Court	Date of conviction	Penalty imposed, including terms of any endorsement/disqualification

10. (a) Was there any period of time in the last four years during which you were disqualified for driving? Yes ☐ No ☐ *

    (b) If YES, give details of the period of disqualification ....

11. (a) Except for spent convictions †, have you ever been convicted on a non-motoring offence? Yes ☐ No ☐ *

(continue overleaf)

* Please tick appropriate box
** "Foreign licence" means a document issued under the law of a country outside the United Kingdom authorising you to drive a motor vehicle in that country.
† The Rehabilitation of Offenders Act 1974, explains when a conviction becomes spent. Further information may be obtained from the Home Office or your legal adviser.

The Department of Transport forms on pages 106-109 inclusive are Crown copyright and are reproduced with the permission of the Controller of Her Majesty's Stationery Office.

(b) If YES give particulars of each offence as follows (continuing on a separate sheet if necessary):

Particulars of offence	Name of Court	Date of conviction	Penalty imposed

12 (a) Are Court proceedings of any kind pending against you?   Yes ☐ No ☐ *

(b) If YES give details ....

....

13 Give names and addresses of two people, not relatives, who have known you for at least one year and who are willing to give a reference as to your character.

1 ....   2 ....

   ....

   ....

   ....

I DECLARE that the above particulars are correct to the best of my knowledge and belief.

I APPLY for registration in the Register of Approved Driving Instructors. I also apply for admission to the written part of the qualifying examination, for which I enclose the appropriate fee. Remittances should be in the form of crossed cheques or Postal Orders made payable to the "Accounting Officer, Department of Transport".

I UNDERSTAND that if I pass the qualifying examinations I am required to pay a registration fee before my name can be entered in the Register.

I UNDERSTAND that the written and practical parts of the qualifying examination and, if I am approved for registration, tests of continued ability and fitness to give instruction, will be conducted during the Supervising Examiner's normal working hours.

I UNDERTAKE that, if approved for registration, I will advise the Registrar of Approved Driving Instructors Department of Transport, 2 Marsham Street, London SW1P 3EB in writing within seven days (i) if I change my address or place of business or employment and (ii) if I am convicted of any offence.

Signed...

Date .... 19.......

* *Please tick appropriate box*

WARNING

An applicant who, for the purpose of securing registration, knowingly makes a false statement or who gives instruction without being in possession of a licence is liable to a fine of up to £500.

Running Your Own Driving School

Department of Transport

ROAD TRAFFIC ACT, 1972 – SECTION 131

Application for a Licence to give Instruction in the Driving of a Motor Car

ADI 3L (revised Aug 1982)

Reference No

Please read these notes carefully before completing the form.

1. *The applicant for a licence should make sure that all the appropriate sections of the form have been completed, and that he has signed section 19. There will be delay in dealing with the application if the form is not properly completed.*

2. *Section 13. There must be at least one Approved Driving Instructor working at an establishment for every licence holder working there. The licence is not valid for use from any other address.*

3. *Section 20 must be signed by the manager or owner of the establishment from which the applicant will be giving instruction under the licence.*

4. *After completion, the form should be sent with the fee to the local Traffic Area Office.*

5. *Information about the licensing arrangements is given in a leaflet (ADI 14) which can be obtained from Traffic Area Offices.*

---

1. Surname .... Mr ☐ Mrs ☐ Miss ☐ Ms ☐ *

   Full christian or other name(s) ....

2. Private address ....

3. Postcode ....

4. Telephone No ....

5. Have you previously applied for

   (a) a licence to give instruction?   Yes ☐ No ☐ *

   (b) registration?   Yes ☐ No ☐ *

6. Do you hold a current substantive (ie not provisional) licence, issued in Great Britain or Northern Ireland, to drive a motor vehicle?   Yes ☐ No ☐ *

7. What is the Driver Number on your driving licence? ☐☐☐

8. Have you held a full driving licence, or a foreign licence**, for periods amounting in the aggregate to at least 4 years in the last 6 years? (after passing driving test any unexpired portion of a provisional licence counts towards required 4 year period).   Yes ☐ No ☐ *

9. (a) Except for spent convictions †, have you ever been convicted of a motoring offence?   Yes ☐ No ☐ *

   (b) If YES, give particulars of each offence as follows (continuing on a separate sheet if necessary)

Particulars of offence	Name of Court	Date of conviction	Penalty imposed, including terms* of any endorsement/disqualification

10. (a) Was there any period of time in the last four years during which you were disqualified for driving?   Yes ☐ No ☐ *

    (b) If YES, give details of the period of disqualification ....

---

* *Please tick appropriate box.*

** *"Foreign licence" means a document issued under the law of a country outside the United Kingdom authorising you to drive a motor vehicle in that country.*

† *The Rehabilitation of Offenders Act 1974 explains when a conviction becomes spent. Further information may be obtained from the Home Office or your legal adviser.*

The Register of Approved Driving Instructors

11 (a) Except for spent convictions†, have you ever been convicted of a non-motoring offence? Yes ☐ No ☐

(b) If YES, give particulars of each offence as follows (continuing on a separate sheet if necessary)

Particulars of offence	Name of Court	Date of conviction	Penalty imposed

12 (a) Are Court proceedings of any kind pending against you? Yes ☐ No ☐

(b) If YES give details

13. Give the name and address of the establishment from which you will give instruction under the licence. (See note 2)

Name _____ Address _____

Tel No _____

14. Have you passed the written part of the qualifying examination for the Register of Approved Driving Instructors within the last 3 years? Yes ☐ No ☐ *

15 (a) Are you waiting to receive the result of an attempt at the written part of the examination? Yes ☐ No ☐ *

(b) If YES, give the examination date and town

16 (a) Have you received an invitation to take the written part in the future? Yes ☐ No ☐ *

(b) If YES, give the examination date and town

17 (a) Are you waiting to receive an invitation to take the written part? Yes ☐ No ☐ *

(b) If YES, give the date of your application

If the answer to each of questions 14, 15, 16 and 17 is "No" you must apply on form AD13 to take the written examination before your application for a licence can be considered.

18 Application

I apply for a licence to give instruction. I understand and accept the conditions under which any licence would be granted and enclose the appropriate fee. Remittance should be in the form of crossed cheques or Postal Orders made payable to the "Accounting Officer, Department of Transport".

19. Declaration

I declare that I have checked the above particulars and that these are correct to the best of my knowledge and belief. I also declare that I have read the Department's leaflet ADI 14.

Signature of Applicant _____ Date _____

20. Declaration and Undertaking by the Manager/Owner of the Establishment Named in Item 13 (See note 3)

I declare that I have read paragraph 15 of leaflet ADI 14 and accept the conditions under which any licence would be granted and that:—

(a) at the date of signing this declaration the ratio is being observed at the address given in item 13;

(b) the grant of a licence will not breach the ratio, and I undertake that at no time during the currency of the licence will the ratio be breached.

Signature of Proprietor/Manager _____ Name in BLOCK LETTERS _____

Date _____ *Please tick appropriate box

> **WARNING**
> An applicant who, for the purpose of securing registration, knowingly makes a false statement or who gives instruction without being in possession of a licence is liable to a fine of up to £500.

109

The qualifying examinations

The examination for entry to the Register consists of a written and a practical part. An applicant must pass the written part before he can take the practical. If he passes the written part of the examination but fails to take or pass the practical within three years, he has to take the written examination again. Details of the examination, including the syllabus, are given below.

The examinations are held at intervals at the following centres:

Written part of the examination. Aberdeen, Birmingham, Bristol, Cambridge, Cardiff, Chatham, Chelmsford, Chester, Eastbourne, Edinburgh, Glasgow, Inverness, Leeds, London, Manchester, Newcastle upon Tyne, Nottingham, Oxford, Plymouth, Preston and Southampton.

Practical part of the examination. Bridgend, Bristol, Cambridge, Chester, Coventry, Darlington, Dartford, Eastbourne, Edinburgh, Epping, Glasgow, Guildford, Hull, Inverness, Isleworth, Leeds, Liverpool, Maidstone, Manchester, Newcastle upon Tyne, Northampton, Norwich, Nottingham, Oxford, Plymouth, Preston, Redhill, Sheffield, Southampton, Taunton, Watford, Wolverhampton and Worcester.

An applicant can choose which centre to attend for the written part of the examination and will get notice of an appointment at least two weeks before the sitting: the notice can be much longer. The results of a written examination are usually sent out within three weeks of the date on which it was held, but may take longer on occasions. Having passed the written part, the applicant will receive details about applying for the practical part of the examination, which can be taken at any centre. Again, at least two weeks' notice of an appointment will be given and the result announced in about three weeks.

The written examination
The written part of the examination for registration consists of a single paper of 100 questions. For every

question a choice of three answers is given. Only one of these three answers is correct. Candidates are required to indicate which is the correct answer by making a mark in the appropriate answer box. The time allowed for the examination is 1½ hours. The examination calls for a high standard of knowledge. The questions will be on all or any of the following subjects:

The principles of road safety generally and their application in particular circumstances;
The techniques of driving a car correctly, courteously and safely, including control of the vehicle, road procedure, recognising hazards and taking proper action, dealing properly with pedestrians and other road users, the use of safety equipment;
The tuition required to instruct a pupil in driving a car, including the items set out above; the correction of the pupil's errors; the manner of the instructor; the relationship between instructor and pupil, and simple vehicle adaptations for disabled drivers;
The *Highway Code* and other matters in the booklet in which it is published;
The Department of Transport booklet DL68 *Your Driving Test*;
The interpretation of the reasons for failure given in document DL24, the *Statement of Failure* to pass the driving test;
Knowledge, adequate to the needs of driving instruction, of the mechanism and design of a car;
The book *Driving* — the official manual issued by the Department of Transport and published by HM Stationery Office.

The practical examination
The practical part of the examination consists of three tests:

Eyesight test
Test of driving technique
Test of instructional ability.

The practical examination lasts altogether about two hours and a candidate must pass all three tests on one and the same occasion.

Eyesight test. The candidate is required to read a motor car number plate at a distance of 90 feet where the letters and figures are 3-1/8 inches high (100 feet for the older 3-1/2 inch symbols), in good daylight, with the aid of glasses if worn. A candidate who is not able to pass the eyesight test will not be eligible to take the other two tests.

Test of driving technique. This test follows the lines of the learner driver test but a higher standard of competence is expected. The candidate must satisfy the examiner that he has an adequate knowledge of the principles of good driving and road safety and that he can apply them in practice. In particular, he must satisfy the examiner on all or any of the following subjects:

Expert handling of the controls;
Use of correct road procedure;
Anticipation of the actions of other road users and taking the appropriate action;
Sound judgement of distance, speed and timing; and
Consideration for the convenience and safety of other road users.

He must also show his ability to perform all or any of the following manoeuvres:

Move away straight ahead or at an angle;
Overtake, meet or cross the path of other vehicles and take an appropriate course;
Turn right-hand and left-hand corners correctly;
Stop the vehicle as in an emergency;
Drive the vehicle backwards and while so doing enter limited openings to the right and to the left; and
Cause the vehicle to face in the opposite direction by the use of forward and reverse gears.

The routes used for the test include busy roads in built-up areas and outside. The applicant will be expected to show that he can make reasonable use of the performance of the vehicle with safety.

Test of instructional ability. The candidate must show his knowledge and ability by giving practical driving instruction to the examiner acting as his pupil, assuming the examiner to be, successively, an absolute beginner, a learner driver with some knowledge (details will be given to the candidate at the time) and one who is at about driving test standard. For each of these three phases, the examiner will nominate one or more of the following subjects as the basis of the instruction:

Explanation of the controls;
Moving off and making normal progress;
Driving the vehicle backwards and while so doing entering limited openings to the right or to the left;
Turning the vehicle round in the road to face in the opposite direction using forward and reverse gears;
Parking close to the kerb, using reverse gear;
Using the mirror, and explaining how to make an emergency stop;
Approaching corners;
Judgement of speed and general road positioning;
Dealing with road junctions;
Dealing with cross-roads;
Meeting, crossing the path of, and overtaking other vehicles; and
Allowing adequate clearance for other road users, and dealing with pedestrian crossings.

The candidate will be assessed on the method, clarity, adequacy and correctness of his instruction, the observation and correction of driving errors committed by his pupil and his general manner.

Cars used for the qualifying examination

The candidate must provide a car for the practical tests. It must be a suitable saloon motor car or estate car in proper condition and capable of the normal performance of vehicles of its type. It must be free from advertisements and signs which might cause other road users to believe that it is being used for the purposes of giving driving instruction, or that it is not being driven by a qualified driver. It should have an orthodox (ie, non-automatic) transmission system, right-hand steering, a readily adjustable driving seat and a seat for a forward-facing front seat passenger. The candidate should have two 'L' plates with him for attachment to the vehicle when required.

The candidate must arrange for the vehicle to be properly insured at the time of taking the practical test. In particular, the car must be fully covered for the following risks for the time during which the examiner is driving:

The examiner's liability for all third party and damage risks, and
His liability to any passenger, including any official passenger.

The Department of Transport will not accept responsibility for risks not covered by insurance, including the loss of any no claims bonus or the cost of repairing minor damage. If the vehicle or the insurance does not meet the above requirements, it will not be possible to proceed with any part of the practical test.

Traffic Area Offices

Northern Traffic Area
Westgate House, Westgate Road,
Newcastle upon Tyne NE1 1TW; 0632 610031

Yorkshire Traffic Area
Hillcrest House, 386 Harehills Lane, Leeds LS9 6NF;
0532 495661

North-Western Traffic Area
Arkwright House, Parsonage Gardens, Deansgate,
Manchester M60 9AN; 061-832 8644

East Midland Traffic Area
Birkbeck House, 14-16 Trinity Square, Nottingham NG1 4BA;
0602 45511

West Midland Traffic Area
Cumberland House, 200 Broad Street, Birmingham B15 1TD;
021-643 5011

Eastern Traffic Area
Terrington House, 13-15 Hills Road, Cambridge CB2 1NP;
0223 358922

South Wales Traffic Area
Caradog House, 1-6 St Andrew's Place, Cardiff,
Glamorgan CF1 3PW; 0222 24801

Western Traffic Area
The Gaunt's House, Denmark Street, Bristol BS1 5DR;
0272 297221

South Eastern Traffic Area
Ivy House, 3 Ivy Terrace, Eastbourne BN21 4QT; 0323 21471

Scottish Traffic Area
24 Torphichen Street, Edinburgh EH3 8HD; 031-229 2515

Metropolitan Traffic Area
PO Box 643, Government Buildings, Bromyard Avenue,
The Vale, Acton, London W3 7AY; 01-743 5522

Personal stories

Margaret
After several years of training and employing a
stream of instructors—only to have them either leave
the profession or set up in businesses of their own—
my husband thought it would be a good idea for me to
join him in his business and become a driving instructor.

At first I was rather reluctant as I thought I had
neither the ability nor the patience to perform the job,
but I agreed to give it a try. For weeks and weeks
(which seemed like years) we spent arduous training
sessions in the car—many times leaving me with a
feeling of total frustration at not being worthy of his
time and tolerance.

Having studied all the appropriate material I presented myself for the Department of Transport ADI
written exam and, much to my amazement, passed.
Now for the real thing!

Working with my husband's supervision under the
trainee licence scheme, I began actually to teach
people to drive and it wasn't long before I received the
appointment for my practical exam. Knees knocking,
I presented myself at the local traffic area offices and
gave what I believed to be an abysmal performance. I

115

taught the supervising examiner how to carry out the 'cockpit drill' and move the car off, then, using all my teaching aids, gave him tuition on dealing with crossroads and how to execute an emergency stop. At the end of the exam I felt utterly drained and burst into tears as soon as I arrived home. This worried my husband somewhat, but three weeks later our fears were allayed—I had passed, first time too!

Since then, and having run my own driving school in the meantime, I've attained a pass rate with my clients of nearly 40 per cent above the national average and over the years I have found the work very interesting and rewarding. It's a job where you meet lots of nice people, most of whom are willing to listen and learn. There is nothing like the feeling of job satisfaction when you return to your car at the end of a client's driving test to be told, 'I've passed, thanks to you.'

Some twelve years ago my husband and I decided that the driving instructor, being an independent type of person, tended to lead a very insular life and being so isolated, never exchanged ideas or views on the profession except during brief encounters with other ADIs in the waiting rooms of test centres. So we set about circulating most of the local instructors with a view to holding a meeting to discuss the formation of an association.

We were rather pleased at the response and an association with more than 30 members was launched, my husband being elected chairman and myself the secretary. Since that time I have held most posts within the association, including chairman, and have been on visits to the Driving Examiners' Training Establishment at Cardington, the Transport and Road Research Laboratory and other relevant establishments, and also to Germany to see how the industry is run over there.

Several years ago, when it was felt that driver education might be taken into the school system, I attended a course at the local college of further education to arm myself with a recognised teaching qualification, the City & Guilds 730 Further Education

Teacher's Certificate, to ensure that my teaching ability would be recognised by education authorities.

In 1978 I decided to attend the National Joint Council of Approved Driving Instructors Organisations tutor training course (staff instructor course for training new entrants to the profession) exchanging views and ideas (and learning a lot!) with some of the top-line instructors of the country. Thus I learned the necessary skills to teach instructors the job.

It was a little embarrassing at first being one of only two women among so many men and it didn't help matters much having my husband as one of the trainers conducting the course! However, he wasn't assigned to 'tutor' me at any stage during the proceedings and at the end of the course I came second in the overall assessments and qualified as an NJC 'tutor'.

I now train my own staff and also work alongside my husband in the training of new instructors from all parts of the country. Most of them have followed the home study course written and devised by us which is aimed at getting would-be instructors through the Department of Transport written exam before attending practical courses.

The job as a whole is very worthwhile and with the changes now in hand to improve standards, I feel it can only get better, both for new instructors, providing they are prepared to work hard, and for the public.

Jamie and Anne

Jamie launched his school in the north of England in 1975 prior to which he had been a lorry driver working at weekends for another driving school. In retrospect he admits that 1975 was a bad time to start up his business because it was in the midst of the petrol crisis caused by the Arab/Israeli war in 1974.

It took well over nine months to build up the business to 'full book' status and it was a financial struggle for the first six to keep going. Jamie's wife Anne had to get a part-time job to help out but it was still difficult as this also meant that there was no one at home to answer the telephone while Jamie was out giving lessons, which seemed to create a vicious circle.

117

An unfortunate accident in August put the vehicle off the road for nearly three weeks while arguments with insurance companies were taking place and difficulty was experienced in obtaining a new body panel for the Japanese car. All this nearly put Jamie off the driving school business altogether, particularly as the accident hadn't been his pupil's fault. It took six months to settle the claim and retrieve his lost earnings from the insurance company, and after this things started to improve. By March the following year, Jamie was doing quite well for himself and business was coming in faster than he could cope with and finances were settling down.

In early 1977 Jamie was so busy that Anne left her job and, after obtaining a second car, she started to work part time under the trainee licence system.

Business continued to improve and, after Anne qualified in September 1977, the couple decided to employ a trainee to work at weekends and in the evenings on a part-time basis. The business then continued in much the same way until 1980 when Jamie felt that they were getting nowhere fast.

After a great deal of heart-searching it was decided to open an office near the town centre. This stretched finances considerably at first but it soon started paying off and enabled them to expand the business and employ other instructors on a full-time basis, the additional vehicles being financed through lease facilities.

In addition to Jamie's car, four others now worked from the office and they have another working in a neighbouring town; Anne now works as the receptionist in the office and still does a few lessons a week to help pay the upkeep of the spare vehicle kept for her personal use, making seven cars in all.

'Competition is tough, and when an instructor left last year to start up his own driving school, I didn't bother replacing him,' said Jamie, when interviewed about his business.

Harry
Harry was a man full of ambition! He sought employment as a trainee driving instructor with one of the

larger schools in the city. While he was training he was often overheard in the local test centres saying how he was going to open a driving school in his own home town and put all the local instructors out of business.

Sure enough, soon after qualifying, Harry began to advertise his services in the local paper. He left his sponsoring school, in debt to them, and began operating his own business with no experience at all. In order to attract business in the shortest possible time, Harry advertised ridiculously low and uneconomical fees and denigrated members of a local professional driving instructors' organisation in the press by accusing them of keeping clients on their books for too long, therefore increasing the number of lessons sold.

Certain sectors of the public are very gullible— those always looking for the 'cheap' bargains—and so, for a while, Harry enjoyed a steady inflow of business. He then began advertising claims such as, 'We have such faith in our ability to get you through the test first time that, in the unlikely event of a failure, we will pay your next test fee.'

This latest advertising campaign seemed to work for a while, so the next plan of action was expansion. To cover this expansion Harry opened an office (increasing overheads even more) and started employing trainee instructors at low rates of pay. This created a 'blind leading the blind' situation with an instructor, with less than two years' experience himself, training newcomers to the job.

The training given was minimal (if any at all) and the required supervision virtually non-existent. Harry was even employing more than the Department of Transport recommended number of trainees per qualified instructors. Needless to say, because of the quality of tuition, clients began failing their driving tests. This meant outgoing costs were rising considerably because of the 'next fee paid' offer, so this offer was soon withdrawn.

Expansion into the surrounding areas was the next idea. In order to operate such a service, Harry was sending his instructors miles away to pick up clients who, in turn, spent virtually their entire lessons driving to the next pick up point, and so on.

By this time, word had been getting around—and bad news always travels faster than good. Clients were leaving for the better services of other instructors in the area, having realised that cheap in the short term works out expensive in the long run.

Many of Harry's staff never even qualified because of their inadequate training and others left one by one to set up in competition. During the entire operation the other schools in the area were greatly affected and, where there is normally a degree of cooperation between instructors, Harry found himself in complete isolation with all the others resenting both him and his methods of operation.

Prices in general within the industry locally were depressed and other schools were forced to operate at less than economic rates in order to compete and there is a limit to how much more can be charged than the cheaper competition before incoming business levels are affected.

The entire affair lasted for about four years, but gradually, due to the imbalance of Harry's expenditure over income and adverse publicity he received, his business began to diminish. Local dealers would no longer supply vehicles for which they knew they would not be paid and even petrol stations were owed money by the ailing business.

Gradually the business returned to its original status—a one-car operation, now working under a different name so as not to be associated with the former school. But the plan didn't work, and insufficient business and personal circumstances forced Harry to leave the area. However, being a very determined person, he returned—yet again under another title. Alas! Not enough people were taken in this time and Harry was forced to join the dole queue.

Robert and Glenda

What influences a person to become a driving instructor? In Robert's case it was a matter of having decided, at a very early age, that one day he would become his own boss. Robert says, 'I had to wait for quite a long time until the most important ingredient came along and that was someone who would share my ideas, enthusiasm and all the worry and anxiety which would almost certainly occur.' He was referring to his wife Glenda, whose patience and restraining influence, he says, has kept their business on a sound footing. Having shared the love of driving for many years, their obvious choice of business seemed to be driving instruction. 'But,' they wondered, 'how does one get to be a driving instructor?'

Robert offered his services to everyone advertising for trainee instructors until he was eventually accepted by a local driving school. That was the start, for him, of a very happy 12 months' association. But Robert decided it was time to move on and establish his own business and so, on a bleak February day in 1978 he went out and bought his first driving school car.

Not having experienced a winter in the driving instruction profession, Robert had not allowed for the one thing which could stop him working for long periods—snow. The winter proved to be the worst for years. However, they survived and as spring advanced business flourished to such an extent that, with some reluctance Glenda decided to train as an instructor and join the business.

They became more ambitious and it seemed they could do no wrong. Expansion was called for and rapidly took place, bringing with it all the pitfalls of employing personnel. Their dining room became more like an office and less like a home, so the next obvious move was to look for office premises. They then reduced the risks involved and formed a limited company, at the same time opening an office, which in turn brought an increase in trade. By this time Robert was fully employed on administration and supervision, having increased the staff to 17 instructors working in an ever-widening area.

Obviously, with so many people to control, Robert and Glenda have their fair share of headaches and worries. They have regrets as well as the satisfaction of success—less time for personal enjoyment, the telephone ringing continually and, possibly their big-

121

gest worry, are they expanding too much to be able to give the personal touch? However, they both agree that there is no going back—they will always go on.

Diversification

John must be the master of diversification. He opened his first driving school, operating from a small office in 1934 just before the introduction of the driving test a year later. This business had to be temporarily shelved during the war years but was reopened in 1949. In 1950 John went into the car hire business with three pre-war vehicles. Nine years later this had expanded to a hire fleet of 170 cars. After opening his first office in 1960, the fleet expanded to 300.

The current recession provides an example of John's versatility and industriousness; his turnover doubled between 1980 and 1982 and he currently has around 25 employees and owns what he describes as a medium-sized fleet.

The secret of John's success is simple—diversification. When he was teaching people to drive he realised that after they passed the test he never saw them again, so he started the car hire business which led to self-drive van hire, mini coaches, private hire, van deliveries and a removal service. Somewhere along the line in the midst of this diversification, John has opened his own vehicle workshop.

He describes himself as a transport consultant and as a result of his worldwide tours and conference attendances one is left with the impression that there is enormous potential for the small man in the field of transport.

Chapter 9

Training for the Examinations

There has recently been much controversy, both in and outside the industry, about the minimum standards set by the examinations, loopholes in the regulations, and the abuse of trainee licences issued for the purpose of gaining practical experience.

In practice there have been two fundamental flaws in the regulations and the method of qualifying which subsequently developed, although the Register probably achieved most of its objectives both for the time it was pioneered and for the foreseeable future. The weaknesses have shown themselves to be in the standards and structure of the qualifying examinations and the conditions relating to the trainee licence, also in the methods of training new instructors which subsequently developed, to meet the above requirements.

There is an obvious link between these factors in that the qualifying standard has largely been responsible for the methods of training new instructors. The standards of both the written examination and the test of driving technique have recently been revised upwards and similar changes are in hand for the test of instructional ability. In addition, new legislation is currently before Parliament to close the loopholes permitting abuse of the trainee licence system.

Until quite recently the driving school industry was unable to agree upon the provision and control of adequate alternatives to the trainee licence scheme. This lack of cooperation between individual instructor associations and their self-interest have been partly responsible for delaying the much needed revision to the regulations and the proposed legislative changes required to encourage the development and provision of professional training facilities for new instructors.

Some degree of unity was achieved as far as training was concerned, at least for a while, through the National Joint Council of Approved Driving Instructors Organisations. This is more commonly known as the NJC. With the support and assistance of the Department of Transport this organisation conducted special courses for experienced instructors wishing to train new entrants to the industry. In order to qualify these 'tutors' had to pass a special high level driving test at the Department of Transport Examiners Training Establishment and undergo a gruelling four-week course to develop their instructional and tutoring skills. There are currently about 32 of these tutors located between Cornwall and Edinburgh.

Persons seeking training should establish the qualifications of those offering courses, as courses are being offered by inexperienced instructors who are unqualified in the field. Even a good instructor does not automatically possess the skills necessary for training new ones, in the same way that good drivers do not always make good driving instructors. Bad habits can be passed on.

The proper training of driving instructors is a specialist task which demands quite different skills from those involved in the teaching of new drivers. Despite this, most new instructors are 'trained' by others with little or no experience of what is involved. In many cases they have only just qualified themselves.

The pass rate of candidates taking the written examination is only 50 per cent with similar problems for the practical. Under the existing system many trainees have required the maximum issue of two licences and sometimes two, three or even more attempts at the practical examination have been made. The cost of trainee licences is currently £45 each and the cost of the practical examinations £60. This can result in examination fees alone costing more than a professional training course.

The new examinations

In 1982 the written examination was updated and the format changed to a multiple choice question paper. At the same time, the standard was raised and this has resulted in a significant drop in the pass rate of candidates currently taking the examination. Candidates should seek proper training before presenting themselves for the examination, the current cost of which is £25.

New standards for the test of driving technique were introduced in November 1983 and while it is too early to see the consequences of this on the pass rate, it is anticipated that there will be a significant drop where candidates do not seek properly structured training.

Changes to the test of instructional ability are imminent and have yet to be announced. They are likely to involve the introduction of a variable syllabus which will affect the instruction given to complete novices and a general tightening of standards throughout. These changes are intended to promote higher standards of professional driving instruction in the future and to encourage new instructors to seek proper training.

New legislation

The Road Traffic (Driving Instruction) Act 1984 should, when implemented, do much to reduce the exploitation of the current system of training new instructors. The Act provides for the examination to be in three parts:

A written examination
A practical test of driving technique
A practical test of instructional ability.

It will require an applicant for a trainee driving instructor's licence to have passed two parts of the qualifying examination, ie, the written examination and the test of driving technique, before being eligible for such a licence.

The regulations may also require applicants for a trainee licence to prove they have undergone a formal training course in instructional techniques before a licence can be granted. (This provision is still under discussion and it could prove difficult to define 'formal training'.)

Methods of qualifying

Although there is no legal requirement for applicants taking the qualifying examinations to seek proper training it obviously makes sense to do so, as they are unlikely to succeed without adequate preparation. There are three different methods (or combinations thereof) by which to acquire the necessary knowledge and skills:

1. Private study
2. A professional training course
3. The trainee licence system.

1. Private study

Whether you intend to take a professional course or obtain practical work experience through the trainee licence system, private home study is an essential element of learning to become a driving instructor. The range of the subject material is too great to cover fully on a conventional course. These are usually between about one and three weeks' duration depending upon whether the written examination has been passed beforehand. Some courses try to include both the theoretical and practical components of the examinations on the same course.

It makes sense to study for the written examination and take it before attending a course of training in preparation for the practical component. Where employment is being sought through a driving school under the trainee licence system, you are more likely to succeed in an application for a job if you have already passed the written component of the qualifying examinations.

A structured home study pack and audio kit for this purpose is available from 'Autodriva' Training Systems, 313 Godfrey Drive, Kirk Hallam, Ilkeston, Derbyshire DE7 4HU; 0602 324499. The pack provides a comprehensive course linked to Department of Transport requirements. The structured guide contains step-by-step instructions and easy exercises to support and reinforce the student's learning through each stage. The kit provides six workbooks and all the other materials required to pass the written examination. It also includes all the background information needed for the practical examination.

2. Professional training courses

Various courses now exist for persons wishing to qualify as professional driving instructors; they vary in length but are usually between one and three weeks' duration. Scheduling of courses tends to be intermittent and they are currently run on a supply and demand basis. Some courses may include both classroom and practical work while others may be entirely practical.

The ratio of trainees to tutors for in-car practical training ranges from one-to-one up to three-to-one. Courses operating on a three-to-one ratio tend to be inefficient, involving excessive observation time of the other participants.

Some courses appear to provide little structured training and the trainees are simply supplied with provisional licence holders to practise on. While a limited amount of this 'real-life' experience is beneficial, it does not normally provide a properly controlled learning environment. This is particularly important in the early stages where the trainee's confidence can be easily destroyed. This type of training also limits the syllabus coverage in addition to being potentially dangerous. Courses of this nature should be avoided.

Courses are sometimes run by ordinary driving schools to supplement regular income. Standards vary considerably from one area to another and courses may be of questionable value. It is essential to establish details of the course content, aims and objectives before any commitment is made. It is equally relevant to establish the experience and qualifications of those responsible for the instruction.

Deciding which type of course is most suited to your needs is not the easiest of tasks. You may require a condensed course to supplement your training under the trainee licence system or you may decide it is better to prepare yourself for the examinations independently. The golden rule is to take your time and consider the alternatives very carefully and remember that the cheapest option can often work out to be the most expensive. The following should be given full consideration when you calculate the cost of training.

- The cost of travel, board and lodgings where time has to be spent away from home
- The price of the home study pack for preparation for the written examination
- The cost of the practical course (also the theory component of practical training where no home study course has been followed)
- The examinations—written £25: practical £60.

To help you select the most appropriate method of training you should look at the way in which the examinations are held by the Department of Transport.

Before you are allowed to apply for the practical examination you must first apply for and pass the written examination. It will take the Department about three weeks to inform you of the result. If this is favourable you can apply immediately for the practical. You should work on the assumption that it will be at least another three to four weeks before you get an appointment for the practical component of the examination.

At best, therefore, there will be six to seven weeks between taking and passing the written examination to attending for the practical. This time factor can cause serious problems of continuity where courses are designed to cover the written and practical examination syllabuses at the same time. There are obvious advantages in selecting a two-part course, the first part of which should be directed to the needs of the written exam and the second to the practical.

A training course syllabus*
The following are examples of objectives which you
should ensure are included on your training course.
On completing the course, *you should be able to
achieve the following:*

1. Administration
1.1 *Objective.* Carry out tasks relating to company
administration and procedures.
Content. Answering telephone enquiries.
Making appointments.
Keeping pupil progress records.
Handling cash receipts.
Vehicle maintenance and service procedures.
Post-accident procedures.
Breakdown procedures.
Keeping trainee instructors' records.
Driving test application/postponement
procedures.
Use of company vehicle for private purposes.

2. Road traffic law regulations
2.1 *Objective.* Demonstrate a knowledge of the legal
responsibilities of a driver, keeper of a motor
vehicle and supervisor of a provisional licence
holder.
Content. Licence and vehicle regulations.
Road traffic offences.
Parking regulations.
Legal responsibilities of a supervising driver.
Conditions of use for provisional driving licence.

3. Mechanical principles and preventative maintenance
3.1 *Objective.* Locate and state the function and principles of
major vehicle systems and components.
Content. Vehicle systems: engine, transmission,
suspension, electrics, brakes, steering, body.
Components: engine cooling, lubrication and
power systems:
(i) clutch, gearbox, differential
(ii) distributor, points, coil, plugs, battery
alternator
(iii) carburettor, fuel pump, exhaust system
(iv) disc and drum brakes
(v) leaf and coil springs, shock absorber
3.2 *Objective.* Identify and report vehicle faults which may
lead to further mechanical damage or the

---

*Acknowledgements to 'Autodriva' Training Systems for
permission to quote this.

	possible rejection for a Department of Transport driving test.
Content.	Oil pressure, temperature, brake, fuel, battery warning lights. Any unusual noises or feel to the vehicle.
	Brakes, lights, steering, tyres, horn, seat belts, windscreen wipers/washers, exhaust.
	Dual accelerator must be inoperative for driving tests.
	Engine tick-over speed should not be excessive when vehicle is used for driving test.

3.3 *Objective.* Carry out simple maintenance tasks.

Content. Vehicle cleanliness.

Daily/weekly driver checks of vehicle.

Changing of bulbs, wheels, plugs, oil, fan belt.

4. Driving theory and practice

4.1 *Objective.* Explain, set and answer questions on, and clarify the rules and content of the Highway Code, the booklet DL68 (*Your Driving Test*) and the Department of Transport's own manual *Driving*.

4.2 *Objective.* Explain the principles of low-risk driving strategies, vehicle dynamics, car control skills and human limitations.

Content. Actual and possible risks.

Reasonable and possible precautions.

Forces inherent in the moving vehicle.

Maximising vehicle stability.

Human attention/perception levels.

Active visual search, hazard recognition skills.

Assessment and decision-making skills.

Effects of fatigue, ill-health, drugs, alcohol and emotional stress.

4.3 *Objective.* Demonstrate the principles of safe driving and the procedures outlined in the Highway Code, booklet DL68 and *Driving*.

Content. Carry out demonstrations under ADI test conditions.

Analysis and demonstration of skills and procedures using key word commentaries.

5. Learning theory and practice

5.1 *Objective.* State the methods by which learning occurs.

Content. Imitations, repetition, association and sympathy.

5.2 *Objective.* Outline the processes involved in the development of driving skills.

Content. Awareness.

Manipulative and decision-making skills.

Analysis of the skills.

Development of the skills.

Teaching from known to unknown, simple to complex and concrete to abstract.

5.3 *Objective.* Explain the pre-driving checks, the function and use of the controls and the instrumentation of the motor car.

Content.
(a) Licence and eyesight checks.
(b) Initial entry checks and 'cockpit' drill.
(c) Precautions and starting the engine.
(d) Hand controls:
   (i) steering wheel and associated controls
   (ii) handbrake—purpose and operation
   (iii) gears—purpose and operation
(e) Foot controls:
   (i) accelerator—purpose
   (ii) brake-purpose
   (iii) foot operation/position for accelerator/brake
   (iv) clutch—purpose and operation
(f) Moving practice in the use, operation and further familiarisation of controls in (c), (d) and (e) above.
(g) Purpose, operation and familiarisation with controls, aids and instruments not included in (c), (d) and (e). (It is not normally necessary to include this material on a first driving lesson.)

5.4 *Objective.* Explain, demonstrate and provide the appropriate practice in the driving skills and procedures outlined in the Highway Code and booklet DL68.

Content. Selection of suitable training routes.
Clarity and accuracy of explanations and route directions.
Clarity and correctness of demonstrations.
Instruction at the correct level for the ability and knowledge of the 'L' driver.
Observations and assessment of the 'L' driver.
General manner and attitude toward the learner, eg, encouraging, sympathetic, firm and tactful.

5.5 *Objective.* Improve the 'L' driver's performance.
Content. Perception of faults.
Identifying the causes of errors.
Assessment of training needs.
Recording and keeping accurate records of progress.

5.6 *Objective.* Make proper use of dual controls and other intervention methods.
Content. Reading road and traffic conditions.
Observations and previous knowledge of the learner.
Verbal intervention.

131

Preparedness to use dual controls.
Preparedness to take control of steering wheel.
Dangers of intervention and the need to inform the learner where and why the dual controls are used.

6. Department of Transport driving test

6.1 *Objective.* Conduct a 'mock' driving test to a realistic standard.

Content. Test structure (the DL68).
Method of assessment.
Driving test documentation (forms D10, DL24/DL25).
Procedures for booking/postponing/cancelling tests.
Construction of Highway Code questions.
Bribery, impersonation, rights of appeal and professional relationship with Department of Transport driving test examiners.

Before attending such a course you should:

- Ensure you can meet the minimum legal requirements outlined for instructors on pages 98-9.
- Ensure the content of both the written and practical components cover the examination syllabus.
- Establish, as far as possible, that the person conducting the course is qualified and experienced in the skills required for training instructors.

Upon being entered on the official Register, you will be fully qualified to:

Work for an employer as a driving instructor
Work on contract for a larger, established school
Establish and run your own driving school.

3. The trainee licence system

The trainee licence scheme is appropriate only to those seeking employment with an established driving school. Provisions are contained within the Road Traffic Act 1972 for new instructors to gain practical experience before taking the qualifying examinations, providing they hold a 'provisional' licence to instruct.

Only two such licences, each of six months' duration, can normally be issued under the regulations.

In theory, these licences are issued to enable new instructors to gain experience after an initial course of training. In practice, however, very little training is given by some of the sponsoring schools and many new instructors have been exploited. Sometimes under this system new instructors are paid very low wages in exchange for the training which, in fact, they rarely receive. Training by some unscrupulous instructors, if given at all, is minimal and consists of little more than sitting in the back of a car for a few days watching the sponsoring instructor perform.

On receipt of the trainee licence from the Department of Transport, some trainees find themselves bundled into a car with real learner drivers of all aptitudes and standards, to discover that they are totally unprepared for what ensues. This is the worst example of the blind leading the blind and can be dangerous and emotionally disturbing for both the new instructors and the learner drivers involved.

Around 50 per cent of instructors training under this system never qualify, which can have severe financial implications in addition to ruined career prospects. Many ultimately lose their jobs because the Department of Transport is normally unable to issue more than two licences.

In contrast, many existing instructors have tried to provide a proper and conscientious course of training under this system. Unfortunately this has left them open to exploitation by persons claiming to require employment when they really intend starting up their own business in competition with their unsuspecting sponsors. This has resulted in many of the most qualified and experienced instructors withdrawing from training new instructors altogether which in turn has led to even lower standards, as training is left to instructors who are themselves inexperienced and who do not always care about a professional approach.

Consider the following carefully before selecting an appropriate method of training:

The cost of the trainee licence, which is currently £45.

The cost of the examinations—written £25; practical £60.

The cost of supervision time for trainee licence holders (20 per cent of instruction time for the first three months' duration of licence must be supervised by the sponsoring instructor).

The cost of supplementary full- or part-time training courses.

You are more likely to succeed in any application for a job as a driving instructor if you have already passed the written component of the qualifying examinations. Home study kits are available for this purpose.

To obtain a trainee licence you must:

- Meet the minimum legal requirements for instructors listed on pages 98-9.
- Find a driving school or qualified driving instructor prepared to sponsor and employ you.
- Be prepared to comply with the requirements of the licence relating to its use, records and supervision.
- Apply to take the written examination.

The Road Traffic (Driving Instruction) Act 1984 will require applicants for trainee licences to pass the written examination and test of driving technique before such licences can be granted.

The issue of this licence will permit you to work only from the establishment named on the licence and under the direct supervision of the named sponsor.

To reduce the chances of exploitation and avoid some of the pitfalls of this system, make some initial commonsense checks which should include:

- Ensuring the school is well established and has a good reputation.
- Ensuring the supervising/sponsoring instructor has at least four to five years of

practical experience in the teaching of new drivers.

- Establishing what the training is to consist of in preparation for:
  (a) The written examination
  (b) The test of driving technique
  (c) The test of instructional ability.
- Establishing how and when the training will be carried out.
- Finding out how much training will be conducted before you start teaching the public and also what this will consist of.
- Finding out what you are to be paid during the training period.
  (It is particularly important to establish clearly how this is to be calculated. Most schools will only pay you for the actual number of lessons you conduct and not for any non-working hours during the working day.)
- Establishing how supervision of your working hours is to be achieved.
- Establishing precisely what the pay will be after you have qualified—use a 40-hour week and a six-month period to calculate how much your training is really costing.

It takes about three weeks for the Department of Transport to issue a trainee licence. The following regime is recommended before taking learner drivers for instruction and this three-week period should be used for it.

1. At least 20 hours of home studies
2. At least six hours' evaluation of driving technique and correction
3. At least 30 hours of basic instructional training in the car by the sponsoring ADI
4. Some experience with a real learner (who is not paying for the lesson)
5. At least 10 hours' observation of an experienced instructor with learners of all standards and aptitudes.

If your prospective employer is not organising all of

this for you, or if the training provisions seem inadequate, you may be able to persuade him to provide you with more. Supplementary courses and home study packs are available to assist schools training new instructors.

Chapter 10

Driving Instructors' Associations

One way in which the individual instructor can help himself play a fuller part in the future of his chosen vocation is to become an active member of a driving school organisation. In both their own interest and the standards of road safety, driving instructors should learn to communicate more freely with each other and also with organisations which share their interests.

The driving school industry as a whole possesses a wealth of information, knowledge and experience which should be shared for the benefit of society and to further road safety. By the same token, the driving school industry must recognise that it is to its own advantage to learn from the knowledge and expertise to be found within other road safety organisations.

Local associations

There are many small local organisations spread throughout the country. Most of these are very active, democratic groups of enterprising instructors formed together to further their common interests and to improve on the quality of services available to the public.

This type of association has much to offer the driving instructor in the form of friendly advice, local seminars, group visits, social functions etc. Meetings tend to be discursive, enthusiastic but generally easy-going and informal affairs. They do have a more serious side however and local views can be expressed through membership of the national associations.

National associations

DIA (Driving Instructors' Association), Lion Green Road, Coulsdon, Surrey CR3 2NL. The largest association for driving instructors in the world, with over 6000 members, it is primarily concerned with the provision of services and of representing the driving instructor's rights in Parliament. It aims to raise the standard of the professional driving instructor and obtain public recognition for his role in raising the standards of driver education. Its policies are guided by management committees and it has a positive attitude towards professionalism. The DIA provides its members with a 60-page bi-monthly magazine.

In association with the AEB (Associated Examining Board) it has developed and announced an examination to provide driving instructors with a higher recognised diploma of professional expertise.

MSA (Motor Schools Association), Atherton House, 12 Tilton Street, London SW6 7LR. Established about 1935, the MSA has its head office in London with branches throughout the country. Its policies are controlled by a management board constituted from regional officers. Its purpose is to act as a trade association, representing both motor schools and driving instructors at parliamentary level. The MSA provides both national and regional magazines for its members and courses are available for classroom instructors and staff tutors on business management and film/video making.

The RAC Register of Instructors, RAC House, Lansdowne Road, Croydon CR9 2JA, was formed when the driving test was introduced in 1935 to establish a recognised qualification for professional driving instructors. Since the compulsory registration of instructors was introduced by the Department of Transport in 1970 the RAC Register has remained as a further qualification to which instructors may aspire. Entry on to this Register is by passing a practical examination, with the submission of satisfactory references. Its purpose is to serve the public and the driving tuition industry by promoting high quality instruction combined with high standards of personal character and professional integrity. The RAC Register provides various services to its members, including a monthly newsletter and quarterly journal, *The Instructor*. It does not regard itself as an alternative to the national driving instructor organisations, which its members are free to join.

NAADI (National Association of Approved Driving Instructors), 138 Wellington Road, Manchester M20 9FM. A democratic organisation formed by the amalgamation of two separate bodies (The Society of Approved Driving Instructors and Driving Instructors Limited) in 1976, whose policies are governed by its own members. Officers are elected annually and a postal ballot is made of members on important issues. It provides members with a bi-monthly newsletter and other services.

Diploma in Driving Instruction

This is a professional qualification for driving instructors and is quite separate from the official requirements of the Department of Transport. It is awarded jointly by the Associated Examining Board and Driving Instructors' Association.

The Associated Examining Board (AEB), Wellington House, Station Road, Aldershot GU11 1BQ, is an international examining board whose chief activity is in administering GCE 'O' and 'A' level examinations.

The Diploma will be awarded to those who possess and demonstrate their professional skills by achieving the required standard. The overall aim of the introduction of this qualification is to improve standards of driver education, but it is clear that it will also enhance the professional status of the driving instruction industry. Instructors who are successful in gaining the diploma will be eligible to benefit from the DIA's special professional services being developed for Diploma holders.

The Diploma in Driving Instruction is essential for two reasons: it enables the public to identify those instructors who provide a professional service, and allows the more proficient instructor to gain respect and recognition for his services. This will benefit business activities and improve career prospects.

The structure and form of the Diploma
The Diploma consists of five separate modules. Each one is complete in itself and tested in a paper of two hours' duration. The modules are:

I Legal Obligations and Regulations
Aims: To promote an understanding of the legal obligations of the driver and enable the instructor to carry out his professional duties within the law. To develop an understanding of the Department of Transport Driving Test administration and regulations.

II Management of a Small Driving School
Aims: To provide an understanding of the financial, administrative and professional skills involved in the operation of a small driving school.

III Vehicle Maintenance and Mechanical Principles
Aims: To develop an awareness of the importance of maintaining motor vehicles in a roadworthy and clean condition and provide a sound basic knowledge of their working principles. To provide an understanding of the instructor's responsibilities concerning maintenance and servicing requirements.

IV Driving Theory, Skills and Procedures
Aims: To develop a thorough knowledge and understanding of the responsibilities of driving and the physical and psychological requirements involved. To develop a general understanding of the attitude, skills and procedures involved in reduced risk driving.

V Instructing: Practices and Procedures
Aims: To provide a basic understanding of the principles of learning as they relate to driving instruction and to assist the instructor to design and develop appropriate courses. To develop valid methods of assessing driving competence and provide a thorough understanding of the objectives, syllabus and meth-

ods of assessment used by Department of Transport Driving Test Examiners.

A certificate for each module is awarded to successful candidates. When all five examination modules are passed, candidates are eligible for the award of the Diploma and should apply to the Driving Instructors' Association for this and details of the professional services for Diploma holders.

A detailed syllabus for each module is available from the DIA (£1.50).

Entering for the Examination

The first examinations are being held in May 1984, then in November 1984, and afterwards at the same times of the year. Candidates may enter one or more modules on either of these occasions.

Entries will be accepted at any registered AEB centre which accepts external candidates; these are mostly colleges of further education. A register of AEB examination centres for this Diploma is available upon application to the AEB (The Secretary General ED/5).

The examination entry fee, payable to the examination centre on registering the entry, is £12.50 per module, plus a basic entry fee of £5 irrespective of how many modules are taken. An examination centre may levy a local fee to cover the administrative work involved and/or invigilation costs.

The results of the examinations will be available in July for the May examination and January for the November examination.

The Board's regulations for GCE examinations as set out in the booklet 'Regulations for Examinations' apply to the Diploma, given that obvious amendments are necessary because the examination takes place at a different time from the GCE examinations. It should be noted in particular that Regulation 9 concerning the conduct of examinations applies to all examinations and that Regulation 8c external (private) candidates—local centres—also applies. (A copy of this booklet is available from the AEB's Publication Department price 50p.)

Studying for the examination

Study materials designed for independent learning have been specially prepared for the Driving Instructors' Association and are available from their headquarters. They are:

The DIA Instructor's Manual. This loose-leaf reference manual provides comprehensive coverage of the five topic areas which are set out in Modules I to V of the Diploma in Driving Instruction and will be updated regularly. Its main author was instrumental in developing the Diploma in Driving Instruction.

The Diploma in Driving Instruction Home Study Pack. This pack prepares students for the examinations leading to the Diploma in Driving Instruction and consists of a structured guide containing five workbooks which relate to the five modules of the Diploma. The study pack will be available from the office of the DIA.

Additional resource material

Driving, The Department of Transport (HMSO).

The Driving Instructor's Handbook, J Miller & N Stacey (Kogan Page).

Guidelines for Driver Instruction, OECD (HMSO).

The Highway Code, HMSO.

Know Your Traffic Signs, The Department of Transport (HMSO).

The New Castrol Book of Car Care, Haynes.

Working for Yourself, G Golzen (Kogan Page).

Reports, issued by The Transport & Road Laboratory.

Appendix

Further reading

Consumer Law for the Small Business,
  Patricia Clayton, Kogan Page 1983.

*Employers' Guide to National Insurance
  Contributions,* DHSS (free).

The Guardian Guide to Running a Small Business,
  4th edn, ed Clive Woodcock, Kogan Page 1984.

Law for the Small Business, 3rd edn,
  Patricia Clayton, Kogan Page 1982.

Setting up a New Business, Maurice Gaffney,
  Department of Industry Small Firms Service
  (free).

The Small Business Guide, Colin Barrow,
  BBC Publications 1982.

Working for Yourself, 6th edn, Godfrey Golzen,
  Kogan Page 1983.

Taking up a Franchise, Godfrey Golzen, Colin
  Barrow and Jackie Severn, Kogan Page 1983.

Be Your Own PR Man, Michael Bland,
  Kogan Page 1983.

Useful addresses

National telephone dialling codes are given, though
local codes may differ.

Local councils and Chambers of Commerce can be
good sources of help and information. Many organisations listed below will have good local offices.

Department of Industry, Small Firms Division
  Ashdown House, 127 Victoria Street,
  London SW1E 6RB; 01-212 8667

The Small Firms Division of the Department of Industry will be glad to advise anyone contemplating setting up their own business. There is no cost for the first consultation and only a minimal cost involved should you require any further consultations, which in the case of a driving school is unlikely. The Department has also established a number of regionally based *Small Firms Centres*; dial 100 and ask for freefone 2444:

London and South Eastern Region
Ebury Bridge House, 2-18 Ebury Bridge Road, London SW1 8QD

South Western Region
5th Floor, The Pithay, Bristol BS1 2NB

Northern Region
22 Newgate Shopping Centre, Newcastle upon Tyne NE1 3EE

North West Region
320-25 Royal Exchange Buildings, St Anne's Square, Manchester M2 7AH

1 Old Hall Street, Liverpool L3 9HJ

Yorkshire and Humberside Region
1 Park Row, City Square, Leeds LS1 5NR

East Midlands Region
48-50 Maid Marian Way, Nottingham NG1 6GF

Northern Ireland Development Agency
Maryfield, 100 Belfast Road, Hollywood, County Down

Local Enterprise Development Unit
Lamont House, Purdy's Lane, Newtownbreda, Belfast BT8 4AR; 0232 691031

Scottish Development Agency
120 Bothwell Street, Glasgow G2 7JP; 041-248 2700

102 Telford Road, Edinburgh EH4 2NP; 031-343 1911

Welsh Development Agency
Treforest Industrial Estate, Pontypridd, Mid Glamorgan CF37 5UT; 0443 852666

Advisory, Conciliation and Arbitration Service (ACAS)
Head Office, Cleland House, Page Street, London SW1P 4ND; 01-222 8020

Alliance of Small Firms and Self-Employed People
42 Vine Road, East Molesey, Surrey KT8 9LF; 01-979 2293

British Insurance Brokers Association
Fountain House, 130 Fenchurch Street, London EC3M 5DJ; 01-623 9043

Council for Small Industries in Rural Areas (CoSIRA)
141 Castle Street, Salisbury, Wiltshire SP1 3TP; 0722 336255

Forum of Private Businesses
Ruskin Rooms, Brewery Lane, Knutsford, Cheshire WA16 0ED; 0565 4467

Health and Safety Commission
Regina House, 259 Old Marylebone Road, London NW1 5RR; 01-723 1262

Health and Safety Executive
25 Chapel Street, London NW1 5DT; 01-262 3277

HM Customs and Excise
VAT Administration Directorate, King's Beam House, Mark Lane, London EC3R 7HE; 01-283 8911

Industrial and Commercial Finance Corporation (ICFC)
91 Waterloo Road, London SE1 8XP; 01-928 7822

National Federation of Self-Employed and Small Businesses
32 St Annes Road West, Lytham St Annes, Lancashire FY8 1NY; 0253 720911

The Office of Fair Trading
Consumer Credit Licensing Branch, Government Building, Bromyard Avenue, The Vale, London W3 7BB; 01-743 0611

Registrar of Companies
Companies House, Crown Way, Maindy, Cardiff CF4 3UZ; 0222 388588

102 George Street, Edinburgh EH2 3DJ;
031-225 5774

43-7 Chichester Street, Belfast BT1 4RJ;
0232 234121

Trade Marks Registry
Patents Office, 25 Southampton Buildings,
London WC2 1AY; 01-405 8721

Department of Transport

Driver Vehicle Licensing Centre
Swansea SA99 1AN; 0792 72134
The Registrar for Approved Driving Instructors
2 Marsham Street, London SW1P 3EB;
01-212 3434
Transport and Road Research Laboratory
Crowthorne, Berkshire; 03446 3131

Driving School Suppliers

Big G Products
Hope Yard, Langhedge Lane, London N18;
01-807 5951, 803 7346
Driving Instructors' Association
Lion Green Road, Coulsdon, Surrey CR3 2NL;
01-668 9822
Driving School Supplies Ltd
227 Ashton Lane, Birmingham B20 3HY;
021-356 7467
Midland Magnetic Signs
20 Baliff Street, Northampton; 0606 30332
Motorschool Aids
Rillbank, 23 Melrose Road, Galashiels TD1 2AT;
0896 2994
Royal Automobile Club
PO Box 100, RAC House, Lansdowne Road,
Croydon, Surrey CR9 2JA; 01-686 2525

Dual Controls

AID (Cables)
Queensdale Works, Queensthorpe Road,
London SE26; 01-778 7055

Big G Products (Rods)
Hope Yard, Langhedge Lane, London N18;
01-807 5951, 803 7346

He-Man Equipment Ltd (Rods)
Bakers Wharf, Millbank Street, Northam,
Southampton SO1 1QH; 0703 26952

P J Dual Controls (Cable) Motorschool Aids
Rillbank, 23 Melrose Road, Galashiels TD1 2AT;
0896 2994

Porter Dual Controls Ltd (Rods)
5 Forval Close, Wandle Way, Mitcham,
Surrey CR4 4NE; 01-640 3200

National Instructor Associations

Driving Instructors' Association
Lion Green Road, Coulsdon, Surrey CR3 2NL;
01-660 3333

Motor Schools Association
Atherton House, 12 Tilton Street,
London SW6 7LR; 01-385 3128, 3589

National Association of Driving Instructors
138 Wellington Road, Manchester M20 9FM;
061-445 4608

Tuition insurance

Driving Instructors' Association
Insurance Department, 39 Castle Street,
Guildford, Surrey; 0483 65124

Alexander Howden (Insurance Brokers Ltd)
Nelson House, Park Road, Timperley,
Altrincham, Cheshire WA14 5AA; 061-969 4422

Oakley Vaughan (Insurance Brokers Ltd)
5 Old Mill Parade, Victoria Road, Romford,
Essex RM1 2HU; 0708 68613

Vehicle hire

Accrodell
4 Vernon Street, Hull HU1 3DP; 0482 223089

Frank E Conway and Co Ltd (Contract Hire)
3A Clifton Square, Lytham, Lancashire FY8 5JT;
0253 737494
Driving Instructors' Association
Lion Green Road, Coulsdon, Surrey CR3 2NL;
01-660 3333
John Jackson (Contract Hire)
28 Hermitage Road, Hitchin, Hertfordshire;
0462 4577 (3 lines) or 04626 3792 (evenings)

Vehicle Recovery Services

Automobile Association
Fanum House, Basingstoke,
Hampshire RG21 2EA; 0256 20123
Eagle Car Recovery Service Ltd
Trumpers Way, London W7 2QA; 01-843 9009
National Breakdown Recovery Club
Bradford, West Yorkshire BD12 0BR;
0274 671299
Royal Automobile Club
PO Box 100, RAC House, Lansdowne Road,
Croydon, Surrey CR9 2JA; 01-686 2525

Driving Instructor Training Establishments

Autodriva Training Systems (Nigel and Margaret Stacey)
313 Godfrey Drive, Kirk Hallam, Ilkeston,
Derbyshire DE7 4HU; 0602 324499
Belle Driver Training Centre (David Bellingham)
24 Star Hill, Rochester, Kent ME1 1XB;
0634 41752
BSM
81-87 Hartfield Road, Wimbledon,
London SW19 3TJ; 01-540 8262
Driving Instructors' Association
Lion Green Road, Coulsdon, Surrey CR3 2NL;
01-660 3333
Driver Training Consultancy (Alan Fleet)
3 Harley Green, Leeds,
West Yorkshire LS13 4PX; 0532 570454

Driving Tuition Centre (Don Hodgkinson)
15 Lower Northern Road, Hedge End,
Southampton SO3 4FN; 048 922104
Drivewell (J H Trewhella)
13 Westborne Heights, Redruth,
Cornwall TR15 2TG; 0209 215124, 216440
Peter Edwards
41 Edinburgh Road, Cambridge CB4 1QR;
0223 359079
Farnworth School of Motoring (Peter W Howard)
100 Manchester Road, Kearsley,
Bolton BL4 8NZ; 0204 72789
Highway Driving School (Malcolm G W Cazaly)
1 Cammo Gardens, Edinburgh EH4 8EJ;
031-339 5388
Millers Motor School (John Miller)
57 North Street, Chichester, West Sussex;
0243 783540
John Milne
121 Marshalswick Lane, St Albans,
Hertfordshire AL1 4UX; 0727 58068
Power Training Centre
47 Church Street, Werrington,
Peterborough PE4 6QB; 0733 71986
Road Teachers Education (Michael Aimable)
70 Wallisdean Avenue, Fareham,
Hampshire PO14 1HS; 0329 23485
Road Transport Industry Training Board
MOTEC, High Ercall, Telford,
Shropshire TF6 6RB; 0952 770441;
course bookings 0952 770831
Claire Simmonds
North Close, Milverton, Taunton, Somerset;
0823 400235
Talbot School of Motoring (Dennis Talbot)
New Bedford Road, Luton; 0582 422144
Wellingborough Driving Academy (Joe Smailes)
45 Broad Green, Wellingborough,
Northamptonshire NN8 4LH; 0933 26682

Fred Williams
  84 Beaumont Avenue, St Albans,
  Hertfordshire AL1 4TP; 0727 47332

For details of your nearest NJC tutor contact:
  John Milne, 121 Marshalswick Lane, St Albans,
  Hertfordshire AL1 4UX; 0727 58068

Index